ADVANCED GAS TURBINE CYCLES

ADVANCED GAS TURBINE CYCLES

J. H. Horlock F.R.Eng., F.R.S.

Whittle Laboratory
Cambridge, U.K.

2003

An imprint of Elsevier Science

AMSTERDAM · BOSTON · HEIDELBERG · LONDON · NEW YORK
OXFORD · PARIS · SAN DIEGO · SAN FRANCISCO · SINGAPORE
SYDNEY · TOKYO

ELSEVIER SCIENCE Ltd
The Boulevard, Langford Lane
Kidlington, Oxford OX5 1GB, UK

First edition 2003

Library of Congress Cataloging in Publication Data
A catalog record from the Library of Congress has been applied for.

British Library Cataloguing in Publication Data
A catalogue record from the British Library has been applied for.

ISBN: 0-08-044273-0

⊚ The paper used in this publication meets the requirements of ANSI/NISO Z39.48-1992 (Permanence of Paper).

Transferred to Digital Printing in 2012

To W.R.H.

CONTENTS

Preface xiii
Notation xvii

Chapter 1.   **A brief review of power generation thermodynamics**   1

1.1.   Introduction   1
1.2.   Criteria for the performance of power plants   4
  1.2.1.   Efficiency of a closed circuit gas turbine plant   4
  1.2.2.   Efficiency of an open circuit gas turbine plant   6
  1.2.3.   Heat rate   7
  1.2.4.   Energy utilisation factor   7
1.3.   Ideal (Carnot) power plant performance   7
1.4.   Limitations of other cycles   8
1.5.   Modifications of gas turbine cycles to achieve higher
              thermal efficiency   9
              References   11

Chapter 2.   **Reversibility and availability**   13

2.1.   Introduction   13
2.2.   Reversibility, availability and exergy   14
  2.2.1.   Flow in the presence of an environment at T_0 (not
              involving chemical reaction)   14
  2.2.2.   Flow with heat transfer at temperature T   16
2.3.   Exergy flux   19
  2.3.1.   Application of the exergy flux equation to a closed cycle   20
  2.3.2.   The relationships between ξ, σ and I^{CR}, I^Q   20
2.4.   The maximum work output in a chemical reaction at T_0   22
2.5.   The adiabatic combustion process   23
2.6.   The work output and rational efficiency of an open circuit
              gas turbine   24
2.7.   A final comment on the use of exergy   26
              References   26

Chapter 3   **Basic gas turbine cycles**   27

3.1.   Introduction   27

3.2.	Air standard cycles (uncooled)	28
3.2.1.	Reversible cycles	28
3.2.1.1.	The reversible simple (Joule–Brayton) cycle, [CHT]$_R$	28
3.2.1.2.	The reversible recuperative cycle [CHTX]$_R$	29
3.2.1.3.	The reversible reheat cycle [CHTHT]$_R$	30
3.2.1.4.	The reversible intercooled cycle [CICHT]$_R$	32
3.2.1.5.	The 'ultimate' gas turbine cycle	32
3.2.2.	Irreversible air standard cycles	33
3.2.2.1.	Component performance	33
3.2.2.2.	The irreversible simple cycle [CHT]$_I$	34
3.2.2.3.	The irreversible recuperative cycle [CHTX]$_I$	37
3.2.3.	Discussion	39
3.3.	The [CBT]$_I$ open circuit plant—a general approach	39
3.4.	Computer calculations for open circuit gas turbines	43
3.4.1.	The [CBT]$_{IG}$ plant	43
3.4.2.	Comparison of several types of gas turbine plants	44
3.5.	Discussion	45
	References	46
Chapter 4.	**Cycle efficiency with turbine cooling (cooling flow rates specified)**	47
4.1.	Introduction	47
4.2.	Air-standard cooled cycles	48
4.2.1.	Cooling of internally reversible cycles	49
4.2.1.1.	Cycle [CHT]$_{RC1}$ with single step cooling	49
4.2.1.2.	Cycle [CHT]$_{RC2}$ with two step cooling	51
4.2.1.3.	Cycle [CHT]$_{RCM}$ with multi-step cooling	52
4.2.1.4.	The turbine exit condition (for reversible cooled cycles)	54
4.2.2.	Cooling of irreversible cycles	55
4.2.2.1.	Cycle with single-step cooling [CHT]$_{IC1}$	55
4.2.2.2.	Efficiency as a function of combustion temperature or rotor inlet temperature (for single-step cooling)	56
4.2.2.3.	Cycle with two step cooling [CHT]$_{IC2}$	58
4.2.2.4.	Cycle with multi-step cooling [CHT]$_{ICM}$	59
4.2.2.5.	Comment	59
4.3.	Open cooling of turbine blade rows—detailed fluid mechanics and thermodynamics	59
4.3.1.	Introduction	59
4.3.2.	The simple approach	61
4.3.2.1.	Change in stagnation enthalpy (or temperature) through an open cooled blade row	61
4.3.2.2.	Change of total pressure through an open cooled blade row	62
4.3.3.	Breakdown of losses in the cooling process	64

4.4.   Cycle calculations with turbine cooling   65
4.5.   Conclusions   68
            References   69

Chapter 5.   Full calculations of plant efficiency   71

5.1.   Introduction   71
5.2.   Cooling flow requirements   71
  5.2.1.   Convective cooling   71
  5.2.2.   Film cooling   72
  5.2.3.   Assumptions for cycle calculations   73
5.3.   Estimates of cooling flow fraction   73
5.4.   Single step cooling   75
5.5.   Multi-stage cooling   75
5.6.   A note on real gas effects   82
5.7.   Other studies of gas turbine plants with turbine cooling   82
5.8.   Exergy calculations   82
5.9.   Conclusions   84
            References   84

Chapter 6.   'Wet' gas turbine plants   85

6.1.   Introduction   85
6.2.   Simple analyses of STIG type plants   85
  6.2.1.   The basic STIG plant   85
  6.2.2.   The recuperative STIG plant   90
6.3.   Simple analyses of EGT type plants   91
  6.3.1.   A discussion of dry recuperative plants with ideal heat
            exchangers   91
  6.3.2.   The simple EGT plant with water injection   93
6.4.   Recent developments   97
  6.4.1.   Developments of the STIG cycle   97
    6.4.1.1. The ISTIG cycle   97
    6.4.1.2. The combined STIG cycle   99
    6.4.1.3. The FAST cycle   99
  6.4.2.   Developments of the EGT cycle   99
    6.4.2.1. The RWI cycle   100
    6.4.2.2. The HAT cycle   100
    6.4.2.3. The REVAP cycle   100
    6.4.2.4. The CHAT cycle   101
    6.4.2.5. The TOPHAT cycle   101
  6.4.3.   Simpler direct water injection cycles   103

6.5.	A discussion of the basic thermodynamics of these developments	103
6.6.	Some detailed parametric studies of wet cycles	105
6.7.	Conclusions	107
	References	107

Chapter 7.   The combined cycle gas turbine (CCGT) 109

7.1.	Introduction	109
7.2.	An ideal combination of cyclic plants	109
7.3.	A combined plant with heat loss between two cyclic plants in series	110
7.4.	The combined cycle gas turbine plant (CCGT)	111
7.4.1.	The exhaust heated (unfired) CCGT	112
7.4.2.	The integrated coal gasification combined cycle plant (IGCC)	114
7.4.3.	The exhaust heated (supplementary fired) CCGT	116
7.5.	The efficiency of an exhaust heated CCGT plant	117
7.5.1.	A parametric calculation	118
7.5.2.	Regenerative feed heating	122
7.6.	The optimum pressure ratio for a CCGT plant	123
7.7.	Reheating in the upper gas turbine cycle	126
7.8.	Discussion and conclusions	128
	References	129

Chapter 8.   Novel gas turbine cycles 131

8.1.	Introduction	131
8.2.	Classification of gas-fired plants using novel cycles	132
8.2.1.	Plants (A) with addition of equipment to remove the carbon dioxide produced in combustion	132
8.2.2.	Plants (B) with modification of the fuel in combustion—chemically reformed gas turbine (CRGT) cycles	133
8.2.3.	Plants (C) using non-carbon fuel (hydrogen)	133
8.2.4.	Plants (D) with modification of the oxidant in combustion	135
8.2.5.	Outline of discussion of novel cycles	135
8.3.	CO_2 removal equipment	136
8.3.1.	The chemical absorption process	136
8.3.2.	The physical absorption process	136
8.4.	Semi-closure	139
8.5.	The chemical reactions involved in various cycles	140
8.5.1.	Complete combustion in a conventional open circuit plant	140
8.5.2.	Thermo-chemical recuperation using steam (steam-TCR)	141
8.5.3.	Partial oxidation	143

8.5.4. Thermo-chemical recuperation using flue gases
(flue gas/TCR) 143
8.5.5. Combustion with recycled flue gas as a carrier 144
8.6. Descriptions of cycles 144
8.6.1. Cycles A with additional removal equipment for carbon
dioxide sequestration 144
8.6.1.1. Direct removal of CO_2 from an existing plant 144
8.6.1.2. Modifications of the cycles of conventional plants using the
semi-closed gas turbine cycle concept................. 146
8.6.2. Cycles B with modification of the fuel in combustion
through thermo-chemical recuperation (TCR)............ 147
8.6.2.1. The steam/TCR cycle............................. 149
8.6.2.2. The flue gas thermo-chemically recuperated (FG/TCR) cycle . 150
8.6.3. Cycles C burning non-carbon fuel (hydrogen) 152
8.6.4. Cycles D with modification of the oxidant in combustion.... 154
8.6.4.1. Partial oxidation cycles......................... 155
8.6.4.2. Plants with combustion modification (full oxidation) 158
8.7. IGCC cycles with CO_2 removal (Cycles E) 160
8.8. Summary 162
References 164

CHAPTER 9. **The gas turbine as a cogeneration** 167
(combined heat and power) plant. ....

9.1. Introduction 167
9.2. Performance criteria for CHP plants 168
9.2.1. Energy utilisation factor 168
9.2.2. Artificial thermal efficiency....................... 170
9.2.3. Fuel energy saving ratio 170
9.3. The unmatched gas turbine CHP plant 173
9.4. Range of operation for a gas turbine CHP plant 174
9.5. Design of gas turbines as cogeneration (CHP) plants 177
9.6. Some practical gas turbine cogeneration plants........... 177
9.6.1. The Beilen CHP plant 177
9.6.2. The Liverpool University CHP plant................. 180
References 181

APPENDIX A. **Derivation of required cooling flows.** 183

A.1. Introduction 183
A.2. Convective cooling only......................... 183
A.3. Film cooling................................. 185
A.4. The cooling efficiency 186

| A.5. | Summary | 186 |
| | References | 187 |

| **APPENDIX B.** | **Economics of gas turbine plants** | 189 |

B.1.	Introduction	189
B.2.	Electricity pricing	189
B.3.	The capital charge factor	190
B.4.	Examples of electricity pricing	191
B.5.	Carbon dioxide production and the effects of a carbon tax	192
	References	194

| **Index** | | 195 |

PREFACE

Many people have described the genius of von Ohain in Germany and Whittle in the United Kingdom, in their parallel inventions of gas turbine jet propulsion; each developed an engine through to first flight. The best account of Whittle's work is his Clayton lecture of 1946 [1]; von Ohain described his work later in [2]. Their major invention was the turbojet engine, rather than the gas turbine, which they both adopted for their new propulsion engines.

Feilden and Hawthorne [3] describe Whittle's early thinking in their excellent biographical memoir on Whittle for the Royal Society.

> "The idea for the turbojet did not come to Whittle suddenly, but over a period of some years: initially while he was a final year flight cadet at RAF Cranwell about 1928; subsequently as a pilot officer in a fighter squadron; and then finally while he was a pupil on a flying instructor's course.... While involved in these duties Whittle continued to think about his ideas for high-speed high altitude flight. One scheme he considered was using a piston engine to drive a blower to produce a jet. He included the possibility of burning extra fuel in the jet pipe but finally had the idea of a gas turbine producing a propelling jet instead of driving a propeller".

But the idea of gas turbine itself can be traced back to a 1791 patent by Barber, who wrote of the basic concept of a heat engine for power generation. Air and gas were to be compressed and burned to produce combustion products; these were to be used to drive a turbine producing a work output. The compressor could be driven independently (along the lines of Whittle's early thoughts) or by the turbine itself if it was producing enough work.

Here lies the crux of the major problem in the early development of the gas turbine. The compressor must be highly efficient – it must use the minimum power to compress the gas; the turbine must also be highly efficient – it must deliver the maximum power if it is to drive the compressor and have power over. With low compressor and turbine efficiency, the plant can only just be self-sustaining – the turbine can drive the compressor but do no more than that.

Stodola in his great book of 1925 [4] describes several gas turbines for power generation, and Whittle spent much time studying this work carefully. Stodola tells how in 1904, two French engineers, Armengaud and Lemâle, built one of the first gas turbines, but it did little more than turn itself over. It appears they used some steam injection and the small work output produced extra compressed air – but not much. The overall efficiency has been estimated at 2–3% and the effective work output at 6–10 kW.

Much later, after several years of development (see Eckardt and Rufli [5]), Brown Boveri produced the first industrial gas turbine in 1939, with an electrical power output of 4 MW. Here the objective of the engineering designer was to develop as much power as possible in the turbine, discharging the final gas at low temperature and velocity; as opposed to the objective in the Whittle patent of 1930, in which any excess energy in the gases at exhaust from the gas generator–the turbine driving the compressor–would be used to produce a high-speed jet capable of propelling an aircraft.

It was the wartime work on the turbojet which provided a new stimulus to the further development of the gas turbine for electric power generation, when many of the aircraft engineers involved in the turbojet work moved over to heavy gas turbine design. But surprisingly it was to be the late twentieth century before the gas turbine became a major force in electrical generation through the big CCGTs (combined cycle gas turbines, using bottoming steam cycles).

This book describes the thermodynamics of gas turbine cycles (although it does touch briefly on the economics of electrical power generation). The strictures of classical thermodynamics require that "cycle" is used only for a heat engine operating in closed form, but the word has come to cover "open circuit" gas turbine plants, receiving "heat" supplied through burning fuel, and eventually discharging the products to the atmosphere (including crucially the carbon dioxide produced in combustion). The search for high gas turbine efficiency has produced many suggestions for variations on the simple "open circuit" plant suggested by Barber, but more recently work has been directed towards gas turbines which produce less CO_2, or at least plants from which the carbon dioxide can be disposed of, subsequent to sequestration.

There are many books on gas turbine theory and performance, notably by Hodge [6], Cohen, Rogers and Saravanamuttoo [7], Kerrebrock [8], and more recently by Walsh and Fletcher [9]; I myself have added two books on combined heat and power and on combined power plants respectively [10,11]. They all range more widely than the basic thermodynamics of gas turbine cycles, and the recent flurry of activity in this field has encouraged me to devote this volume to cycles alone. But the remaining breadth of gas turbine cycles proposed for power generation has led me to exclude from this volume the coupling of the gas turbine with propulsion. I was also influenced in this decision by the existence of several good books on aircraft propulsion, notably by Zucrow [12], Hill and Peterson [13]; and more recently my friend Dr Nicholas Cumpsty, Chief Technologist of Rolls Royce, plc, has written an excellent book on "Jet Propulsion" [14].

I first became interested in the subject of cycles when I went on sabbatical leave to MIT, from Cambridge England to Cambridge Mass. There I was asked by the Director of the Gas Turbine Laboratory, Professor E.S.Taylor, to take over his class on gas turbine cycles for the year. The established text for this course consisted of a beautiful set of notes on cycles by Professor (Sir) William Hawthorne, who had been a member of Whittle's team. Hawthorne's notes remain the best starting point for the subject and I have called upon them here, particularly in the early part of Chapter 3.

Hawthorne taught me the power of temperature-entropy diagram in the study of cycles, particularly in his discussion of "air standard" cycles–assuming the working fluid to be a perfect gas, with constant specific heats. It is interesting that Whittle wrote in his later book [15] that he himself "never found the (T,s diagram) to be useful", although he had a profound understanding of the basic thermodynamics of gas turbine cycles. For he also wrote

"When in jet engine design, greater accuracy was necessary for detail design, I worked in pressure ratios, used $\gamma = 1.4$ for compression and $\gamma = 1.3$ for expansion and assumed specific heats for combustion and expansion corresponding to the temperature range concerned. I also allowed for the increase in mass flow in expansion due to fuel addition (in the range 1.5–2%). The results, despite guesswork involved in many of the assumptions, amply justified these methods to the point where I was once rash enough to declare that jet engine design has become an exact science". Whittle's modifications of air standard cycle analysis are developed further in the later parts of Chapter 3.

Hawthorne eventually wrote up his MIT notes for a paper with his research student, Graham de Vahl Davis [16], but it is really Will Hawthorne who should have written this book. So I dedicate it to him, one of several great engineering teachers, including Keenan, Taylor and Shapiro, who graced the mechanical engineering department at MIT when I was there as a young assistant professor.

My subsequent interest in gas turbines has come mainly from a happy consulting arrangement with Rolls Royce, plc and the many excellent engineers I have worked with there, including particularly Messrs.Wilde, Scrivener, Miller, Hill and Ruffles. The Company remains at the forefront of gas turbine engineering.

I must express my appreciation to many colleagues in the Whittle Laboratory of the Engineering Department at Cambridge University. In particular I am grateful to Professor John Young who readily made available to me his computer code for "real gas" cycle calculations; and to Professors Cumpsty and Denton for their kindness in extending to me the hospitality of the Whittle Laboratory after I retired as Vice-Chancellor of the Open University. It is a stimulating academic environment.

I am also indebted to many friends who have read chapters in this book including John Young, Roger Wilcock, Eric Curtis, Alex White (all of the Cambridge Engineeering Department), Abhijit Guha (of Bristol University), Pericles Pilidis (of Cranfield University) and Giampaolo Manfrida (of Florence University). They have made many suggestions and pointed out several errors, but the responsibility for any remaining mistakes must be mine.

Mrs Lorraine Baker has helped me greatly with accurate typing of several of the chapters, and my friend John Stafford, of Compu-Doc (silsoe-solutions) has provided invaluable help in keeping my computer operational and giving me many tips on preparing the material. My publishing editor, Keith Lambert has been both helpful and encouraging.

Finally I must thank my wife Sheila, for putting up with my enforced isolation once again to write yet another book.

J. H. Horlock
Cambridge, June 2002

REFERENCES

[1] Whittle, Sir Frank. (1945), The early history of the Whittle jet propulsion engine, Proc. Inst. Mech. Engrs. 152, 419–435.
[2] von Ohain, H. (1979), The Evolution and Future of Aero-propulsion Systems. 40 Years of Jet Engine Progress. W.J. Boyne, and D.S. Lopez, (ed.), National Air and Space Museum, Washington DC.

[3] Feilden, G.B.R. and Hawthorne, W.R., Sir Frank Whittle, O.M. K.B.E. (1998) Biological Memoirs of the Royal Society, 435–452.

[4] Stodola, A. (1924), Steam and Gas Turbines. McGraw Hill, New York.

[5] Eckardt, D. and Rufli, P. (2000), ABB/BBC Gas Turbines – A Record of Historic Firsts, ASME Turbo-Expo 2000 Paper TE00 A10.

[6] Hodge, J. (1955), Cycles and Performance Estimation. Butterworths, London.

[7] Cohen, H., Rogers, G.F.C. and Saravanamuttoo, H.I.H. (1996), Gas Turbine Theory. Longman, 4th edn.

[8] Kerrebrock, J. (1992), Aircraft Engines and Gas Turbines. MIT Press.

[9] Walsh, P.P. and Fletcher, P. (1998), Gas Turbine Performance. Blackwell Science, Oxford.

[10] Horlock, J.H. (1987), Cogeneration - Combined Heat and Power Plants. Pergamon, 2nd edn, Krieger, Malabar, Florida, 1997.

[11] Horlock, J.H. (1992), Combined Power Plants. Pergamon, 2nd edn, Krieger, Melbourne, USA, 2002.

[12] Zucrow, M.J. (1958), Aircraft and Missile Propulsion. John Wiley, New York.

[13] Hill, P.G. and Peterson, C.R. (1992), Mechanics and Thermodynamics of Propulsion. MIT Press, 2nd edn.

[14] Cumpsty, N.A. (1997), Jet Propulsion. Cambridge University Press.

[15] Whittle, Sir Frank. (1981), Gas Turbine Aero-Themodynamics. Pergamon Press, Oxford.

[16] Hawthorne, W. R., and Davis, G. de V. (1956), Calculating gas turbine performance. Engng. 181, 361–367.

The author is grateful to the following for permission to reproduce the figures listed below.

Pergamon Press, Oxford, UK: Figs. 1.2, 1.3, 9.7 and 9.8

Krieger Publishing Company, Melbourne, Florida, USA: Figs. 1.4, 1.7, 1.8, 2.1, 2.2, 2.3, 2.4, 2.5, 7.3, 7.5, 7.6, 9.5.

American Society of Mechanical Engineers: Figs. 4.1, 4.2, 4.3, 4.4, 4.5, 4.6, 4.7, 4.11, 4.12, 5.4, 5.6, 5.9, 5.10, 5.11, 6.1, 6.8, 6.9, 6.10, 6.12, 6.14, 6.18, 6.19, 6.20, 7.4, 7.7, 7.11, 8.1, 8.2, 8.6, 8.7, 8.13, 8.14, 8.16, 8.17, 8.18, 8.19, 8.20, 8.24, 8.25, 8.26, 8.27, 8.28, A.1, B.1, B.2, B.3.

Council of the Institution of Mechanical Engineers: Figs. 3.8, B.4, 7.9, 7.10.

Princeton University: Figs. 6.2, 6.3, 6.4, 8.11, 8.12.

Pearson Education Limited: Fig. 3.12.

Brown Boveri Company Ltd, Baden, Switzerland: Fig. 7.8.

International Journal of Applied Thermodynamics: Figs. 8.8, 8.23

NOTATION

Note: Lower case symbols for properties represent specific quantities (i.e. per unit mass)

Symbol	Meaning	Typical Units
A	area	m^2
b, B	steady flow availability	kJ/kg, kJ
B	Biot number	$(-)$
C	capital cost	£, $
c_p	specific heat capacity, at constant pressure	kJ/kg K
$[CV]_0$	calorific value at temperature T_0	kJ/kg
d_h	hydraulic diameter	m
e, E	exergy	kJ/kg, kJ
E^Q	work potential of heat transferred thermal exery	kJ
EUF	energy utilisation factor	$(-)$
f	fuel/air ratio; also friction factor	$(-);(-)$
F	fuel energy supplied	kJ
g, G	Gibbs function	kJ/kg, kJ
h, H	enthalpy	kJ/kg, kJ
h	heat transfer coefficient	kW/m^2K
H	plant utilisation	h/year
i	interest or discount rate	$(-)$
I	lost work due to irreversibility (total)	kJ
I^{CR}	lost work due to internal irreversibility	kJ
I^Q	lost work due to heat transfer to the atmosphere	kJ
L	blade length	m
m	mass fraction (e.g. of main steam flow)	$(-)$
M	Mass flow; also fuel cost per annum; also molecular weight: also Mach number	kg/s; £, $ p.a.; $(-)$; $(-)$
n, n'	Ratio of air and gas specific heats, $(c_{pa})/(c_{pg})$	$(-)$
NDCW	non-dimensional compressor work	$(-)$
NDTW	non-dimensional turbine work	$(-)$
NDNW	non-dimensional net work	$(-)$
NDHT	non-dimensional heat supplied	$(-)$
N	plant life	years
OM	annual operational maintenance costs	£, $ p.a.
p	pressure	N/m^2
P	electricity cost per year	£, $ p.a.
q, Q	heat supplied or rejected	kJ/kg, kJ
r	pressure ratio	$(-)$
R	gas constant	kJ/kg K
$\bar{R}$	universal gas constant	kJ/kmol/K
S	fuel costs per unit mass; also steam to air ratio	£, $/kg, $(-)$
s, S	entropy	kJ/kg K, kJ/K
S_t	Stanton number	$(-)$
t	time; also thermal barrier thickness	s, m
T	temperature	°C, K
V	velocity	m/s
w, W	specific work output, work output	kJ/kg, kJ

(*continued on next page*)

(*continued*)

Symbol	Meaning	Typical Units
w^+, W^+	temperature difference ratios in heat transfer	$(-),(-)$
x	isentropic temperature ratio	$(-)$
y	velocity ratio	$(-)$
z	polytropic expansion index	$(-)$
$A, B, C, D, E,$ $F, K K'$	constants defined in text	various
α	proportions of capital cost	$(-)$
α	$= \eta_c \eta_t \theta$	$(-)$
β	$= 1 + \eta_c (\theta - 1)$; also capital cost factor	$(-)$
γ	$= c_p/c_v$	$(-)$
δ	loss parameter	$(-)$
ε	heat exchanger effectiveness; also quantity defined in eqn. [4.24]	$(-)$
ζ	cost of fuel per unit of energy	£, $/kWh
η	efficiency – see note below	$(-)$
θ	ratio of maximum to minimum temperature	$(-)$
λ	area ratio in heat transfer; also CO_2 performance parameter	$(-)$, kg/kWh
μ	scaling factor on steam entropy, ratio of mass flows in combined cycle (lower to upper)	$(-)$
ν	non-dimensional heat supplied (ν_s) or heat unused (ν_{UN})	$(-)$
μ, ξ, σ, τ	parameters in cycle analysis	$(-)$
ρ	density	kg/m^3
τ	T_{min}/T_{max}; also corporate tax rate	$(-)$
ψ	cooling air mass flow fraction	$(-)$
ϕ	temperature function, $\int_0^T \frac{c_p dT}{T}$; also turbine stage loading coefficient	kJ/kg K, $(-)$
σ	expansion index defined in text	$(-)$
κ	constant in expression for stagnation pressure loss	$(-)$

Subscripts

$a, a', b, b', c,$ d, e, e', f, f'	states in steam cycle
a	air
A	relating to heat rejection; artificial efficiency
bl	blade (temperature)
B	boiler; relating to heat supply
c	cooling air
cot	combustion (temperature)
C	compressor (isentropic efficiency)
CAR	Carnot cycle
CC	combustion chamber (efficiency or loss)
CP	combined plant (general)
CG	cogeneration plant
CS	control surface
CV	control volume
d	debt
dp	dewpoint

(*continued*)

Symbol	Meaning	Typical Units
D	demand	
e	maximum efficiency; also equity; also external	
E	electrical (unit price); also exit from turbine, and from first turbine stage	
f	fuel	
g, G	gas	
H	higher (upper, topping), relating to heat supply, work output	
HL	between high and lower plants	
HR	rejection from higher plant	
JB	Joule-Brayton cycle	
i	inlet	
IJB	irreversible Joule-Brayton cycle	
k	product gas component; also year number ($k= 1, 2, \ldots$)	
L	lower (bottoming), relating to heat supply, work output	
LR	rejection from lower plant	
max	maximum	
min	minimum	
m	mixture	
NU	non-useful (heat rejection)	
o	outlet	
O	overall (efficiency)	
p	polytropic (efficiency)	
P	product of combustion	
P'	product of supplementary combustion	
rit	rotor inlet temperature	
R	rational; also reactants	
REV	reversible (process)	
s	steam; also state after isentropic compression or expansion; also surface area (A_s)	
S	state at entry to stack; also supplementary heating	
T	turbine (isentropic efficiency)	
U	useful (heat delivered)	
w	water; also maximum specific work	
x	cross-sectional flow area (A_x)	
X, Y	states leaving heat exchanger; also states at entry and exit from component	
$1, 1', 2, 2',$ $3, 3', 4, 4', \ldots\ldots$	miscellaneous, referring to gas states	
0	conceptual environment (ambient state); also stagnation pressure	

Superscripts

CR	referring to internal irreversibility	
Q	referring to thermal exergy (associated with heat transfer); also to lost work due to external irreversibility associated with heat transfer	
· (e.g $\dot{M}, \dot{Q}, \dot{W}$)	rate of (mass flow, heat supply, work output, etc)	
$'$ (e.g. η')	new or changed value (e.g. of efficiency)	

(*continued on next page*)

(*continued*)

Symbol	Meaning	Typical Units
$'$ (e.g. a', b', $1'$, $2'$, $3'$, $4'$)	states in feed heating train, in reheating or intercooling	
$^-$ (e.g. $\bar{T}$)	mean or averaged (e.g. temperature)	

Note on efficiencies

η is used for **thermal** efficiency of a closed cycle, but sometimes with a subscript (e.g. η_H for thermal efficiency of a higher cycle); η_O is used for (**arbitrary**) **overall** efficiency of a plant.

A list of efficiencies is given below.

Plant Thermal Efficiencies η

η_H	higher cycle
η_L	lower cycle
η_{CP}	combined cycle
η_{CG}	cogeneration plant
η_{CAR}	Carnot cycle

Plant (Arbitrary) Overall Efficiencies η_O

$(\eta_O)_H$	higher plant
$(\eta_O)_{CP}$	combined plant
$(\eta_O)_L$	lower plant

Rational Efficiencies η_R

Component Efficiencies

η_B	boiler
η_C	compressor, isentropic
η_T	turbine, isentropic
η_p	polytropic

Cycle Descriptions

The nomenclature originally introduced by Hawthorne and Davis is followed, in which compressor, heater, turbine and heat exchanger are denoted by C, H, T and X respectively and subscripts R and I indicate reversible and irreversible. For the open cycle the heater is replaced by a burner, B. In addition subscripts U and C refer to uncooled and cooled turbines in a cycle and subscripts 1, 2, ... indicate the number of cooling steps. Thus, for example $[CBTX]_{IC2}$ indicates an open irreversible regenerative cycle with two steps of turbine cooling.

Chapter 1

A BRIEF REVIEW OF POWER GENERATION THERMODYNAMICS

1.1. Introduction

A conventional power plant receiving fuel energy (F), producing work (W) and rejecting heat (Q_A) to a sink at low temperature is shown in Fig. 1.1 as a block diagram. The objective is to achieve the least fuel input for a given work output as this will be economically beneficial in the operation of the power plant, thereby minimising the fuel costs. However, the capital cost of achieving high efficiency has to be assessed and balanced against the resulting saving in fuel costs.

The discussion here is restricted to plants in which the flow is steady, since virtually all the plants (and their components) with which the book is concerned have a steady flow.

It is important first to distinguish between a closed cyclic power plant (or heat engine) and an open circuit power plant. In the former, fluid passes continuously round a closed circuit, through a thermodynamic cycle in which heat (Q_B) is received from a source at a high temperature, heat (Q_A) is rejected to a sink at low temperature and work output (W) is delivered, usually to drive an electric generator.

Fig. 1.2 shows a gas turbine power plant operating on a closed circuit. The dotted chain control surface (Y) surrounds a cyclic gas turbine power plant (or cyclic heat engine) through which air or gas circulates, and the combustion chamber is located within the second open control surface (Z). Heat Q_B is transferred from Z to Y, and heat Q_A is rejected from Y. The two control volumes form a complete power plant.

Usually, a gas turbine plant operates on 'open circuit', with internal combustion (Fig. 1.3). Air and fuel pass across the single control surface into the compressor and combustion chamber, respectively, and the combustion products leave the control surface after expansion through the turbine. The open circuit plant cannot be said to operate on a thermodynamic cycle; however, its performance is often assessed by treating it as equivalent to a closed cyclic power plant, but care must be taken in such an approach.

The Joule–Brayton (JB) constant pressure closed cycle is the basis of the cyclic gas turbine power plant, with steady flow of air (or gas) through a compressor, heater, turbine, cooler within a closed circuit (Fig. 1.4). The turbine drives the compressor and a generator delivering the electrical power, heat is supplied at a constant pressure and is also rejected at constant pressure. The temperature–entropy diagram for this cycle is also

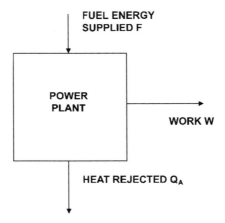

Fig. 1.1. Basic power plant.

shown in the figure. The many variations of this basic cycle form the subject of this volume.

An important field of study for power plants is that of the '*combined plant*' [1]. A broad definition of the combined power plant (Fig. 1.5) is one in which a higher (upper or topping) thermodynamic cycle produces power, but part or all of its heat rejection is used in supplying heat to a 'lower' or bottoming cycle. The 'upper' plant is frequently an open circuit gas turbine while the 'lower' plant is a closed circuit steam turbine; together they form a *combined cycle gas turbine* (CCGT) plant.

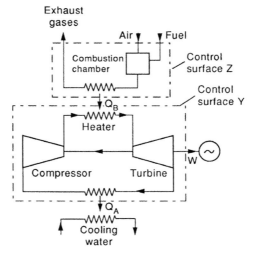

Fig. 1.2. Closed circuit gas turbine plant (after Haywood [3]).

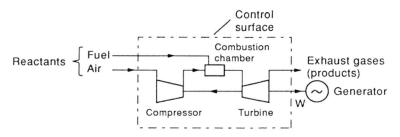

Fig. 1.3. Open circuit gas turbine plant (after Haywood [3]).

The objective of combining two power plants in this way is to obtain greater work output for a given supply of heat or fuel energy. This is achieved by converting some of the heat rejected by the upper plant into extra work in the lower plant.

The term '*cogeneration*' is sometimes used to describe a combined power plant, but it is better used for a *combined heat and power* (CHP) plant such as the one shown in Fig. 1.6 (see Ref. [2] for a detailed discussion on CHP plants). Now the fuel energy is converted partly into (electrical) work (W) and partly into useful heat (Q_U) at a low temperature, but higher than ambient. The non-useful heat rejected is Q_{NU}.

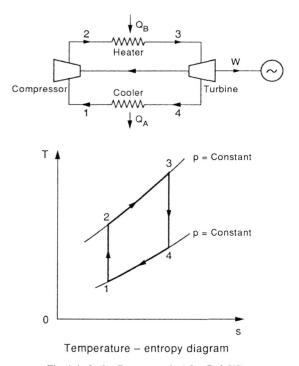

Temperature – entropy diagram

Fig. 1.4. Joule–Brayton cycle (after Ref. [1]).

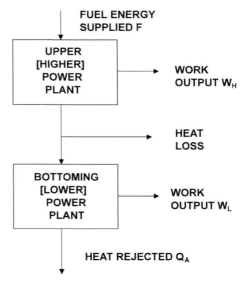

Fig. 1.5. Combined power plant.

1.2. Criteria for the performance of power plants

1.2.1. *Efficiency of a closed circuit gas turbine plant*

For a cyclic gas turbine plant in which fluid is circulated continuously within the plant (e.g. the plant enclosed within the control surface *Y* in Fig. 1.2), one criterion of performance

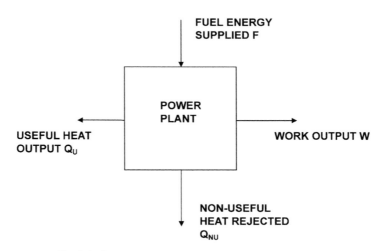

Fig. 1.6. Cogeneration plant (combined heat and power plant).

is simply the thermal or cycle efficiency,

$$\eta \equiv \frac{W}{Q_B}, \tag{1.1}$$

where W is the net work output and Q_B is the heat supplied. W and Q_B may be measured for a mass of fluid (M) that circulates over a given period of time. Thus, the efficiency may also be expressed in terms of the *power* output ($\dot{W}$) and the *rate* of heat transfer ($\dot{Q}_B$),

$$\eta = \frac{\dot{W}}{\dot{Q}_B}, \tag{1.2}$$

and this formulation is more convenient for a steady flow cycle. In most of the thermodynamic analyses in this book, we shall work in terms of W, Q_B and mass flow M (all measured over a period of time), rather than in terms of the rates $\dot{W}$, $\dot{Q}_B$ and $\dot{M}$ (we call M a mass flow and $\dot{M}$ a mass flow rate).

The heat supply to the cyclic gas turbine power plant of Fig. 1.2 comes from the control surface Z. Within this second control surface, a steady-flow heating device is supplied with reactants (fuel and air) and it discharges the products of combustion. We may define a second efficiency for the 'heating device' (or boiler) efficiency,

$$\eta_B \equiv \frac{Q_B}{F} = \frac{Q_B}{M_f[CV]_0}. \tag{1.3}$$

Q_B is the heat transfer from Z to the closed cycle within control surface Y, which occurs during the time interval that M_f, the mass of fuel, is supplied; and $[CV]_0$ is its calorific value per unit mass of fuel for the ambient temperature (T_0) at which the reactants enter. $F = M_f[CV]_0$ is equal to the heat (Q_0) that would be transferred from Z if the products were to leave the control surface at the entry temperature of the reactants, taken as the temperature of the environment, T_0. Fig. 1.7 illustrates the definition of calorific value,

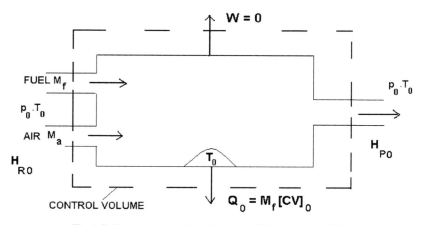

Fig. 1.7. Determination of calorific value $[CV]_0$ (after Ref. [2]).

where Q_0 is equal to $M_f[CV]_0 = [-\Delta H_0] = H_{R0} - H_{P0}$, the change in enthalpy from reactants to products, at the temperature of the environment.

The *overall* efficiency of the entire gas turbine plant, including the cyclic gas turbine power plant (within Y) and the heating device (within Z), is given by

$$\eta_O \equiv \frac{W}{F} = \left(\frac{W}{Q_B}\right)\left(\frac{Q_B}{F}\right) = \eta\eta_B. \tag{1.4}$$

The subscript O now distinguishes the overall efficiency from the thermal efficiency.

1.2.2. *Efficiency of an open circuit gas turbine plant*

For an open circuit (non-cyclic) gas turbine plant (Fig. 1.3) a different criterion of performance is sometimes used—the *rational efficiency* (η_R). This is defined as the ratio of the actual work output to the maximum (reversible) work output that can be achieved between the reactants, each at pressure (p_0) and temperature (T_0) of the environment, and products each at the same p_0, T_0. Thus

$$\eta_R \equiv \frac{W}{W_{REV}} \tag{1.5a}$$

$$= \frac{W}{[-\Delta G_0]}, \tag{1.5b}$$

where $[-\Delta G_0] = G_{R0} - G_{P0}$ is the change in Gibbs function (from reactants to products). (The Gibbs function is $G \equiv H - TS$, where H is the enthalpy and S the entropy.)

$[-\Delta G_0]$ is not readily determinable, but for many reactions $[-\Delta H_0]$ is numerically almost the same as $[-\Delta G_0]$. Thus the rational efficiency of the plant is frequently approximated to

$$\eta_R \approx \frac{W}{[-\Delta H_0]} = \frac{W}{M_f[CV]_0} = \frac{W}{F}, \tag{1.6}$$

where $[-\Delta H_0] = H_{R0} - H_{P0}$. Haywood [3] prefers to call this the (*arbitrary*) *overall efficiency*, implying a parallel with η_O of Eq. (1.4).

Many preliminary analyses of gas turbines are based on the assumption of a closed 'air standard' cyclic plant, and for such analyses the use of η as a thermal efficiency is entirely correct (as discussed in the early part of Chapter 3 of this book). But most practical gas turbines are of the open type and the rational efficiency should strictly be used, or at least its approximate form, the arbitrary overall efficiency η_O. We have followed this practice in the latter part of Chapter 3 and subsequent chapters; even though some engineers consider this differentiation to be a somewhat pedantic point and many authors refer to η_O as a thermal efficiency (or sometimes the 'lower heating value thermal efficiency').

1.2.3. Heat rate

As an alternative to the thermal or cycle efficiency of Eq. (1.1), the cyclic heat rate (the ratio of heat supply rate to power output) is sometimes used:

$$\text{Heat rate} \equiv \frac{\dot{Q}_B}{\dot{W}} = \frac{Q_B}{W}. \tag{1.7}$$

This is the inverse of the closed cycle thermal efficiency, when Q_B and W are expressed in the same units.

But a 'heat rate' based on the energy supplied in the fuel is often used. It is then defined as

$$\text{Heat rate} \equiv \frac{M_f[CV]_0}{W} = \frac{F}{W}, \tag{1.8}$$

which is the inverse of the (arbitrary) overall efficiency of the open circuit plant, as defined in Eq. (1.6).

1.2.4. Energy utilisation factor

For a gas turbine operating as a combined heat and power plant, the 'energy utilisation factor' (EUF) is a better criterion of performance than the thermal efficiency. It is defined as the ratio of work output (W) plus useful heat output (Q_U) to the fuel energy supplied (F),

$$\text{EUF} = \frac{W + Q_U}{F}, \tag{1.9}$$

and this is developed further in Chapter 9.

1.3. Ideal (Carnot) power plant performance

The second law of thermodynamics may be used to show that a cyclic heat power plant (or cyclic heat engine) achieves maximum efficiency by operating on a reversible cycle called the Carnot cycle for a given (maximum) temperature of supply (T_{max}) and given (minimum) temperature of heat rejection (T_{min}). Such a Carnot power plant receives all its heat (Q_B) at the maximum temperature (i.e. $T_B = T_{max}$) and rejects all its heat (Q_A) at the minimum temperature (i.e. $T_A = T_{min}$); the other processes are reversible and adiabatic and therefore isentropic (see the temperature–entropy diagram of Fig. 1.8). Its thermal efficiency is

$$\eta_{CAR} = \frac{W}{Q_B} = \frac{Q_B - Q_A}{Q_B} = \frac{T_{max}\Delta s - T_{min}\Delta s}{T_{max}\Delta s} = (T_{max} - T_{min})/T_{max}. \tag{1.10}$$

Clearly raising T_{max} and lowering T_{min} will lead to higher Carnot efficiency.

The Carnot engine (or cyclic power plant) is a useful hypothetical device in the study of the thermodynamics of gas turbine cycles, for it provides a measure of the best performance that can be achieved under the given boundary conditions of temperature.

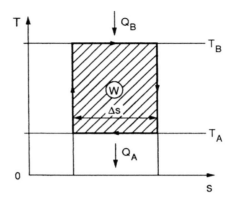

Fig. 1.8. Temperature–entropy diagram for a Carnot cycle (after Ref. [1]).

It has three features which give it maximum thermal efficiency:
(i)   all processes involved are reversible;
(ii) all heat is supplied at the maximum (specified) temperature (T_{max});
(iii) all heat is rejected at the lowest (specified) temperature (T_{min}).
    In his search for high efficiency, the designer of a gas turbine power plant will attempt
to emulate these features of the Carnot cycle.

1.4. Limitations of other cycles

    Conventional gas turbine cycles do not achieve Carnot efficiency because they do not
match these features, and there exist
(i)   'external irreversibilities' with the actual (variable) temperature of heat supply being
      less than T_{max} and the actual (variable) temperature of heat rejection being greater
      than T_{min};
(ii) 'internal irreversibilities' within the cycle.
    Following Caputo [4], we define mean temperatures of heat supply and rejection as

$$\bar{T}_B \equiv \frac{Q_B}{\int \frac{dQ_B}{T}}, \qquad \bar{T}_A \equiv \frac{Q_A}{\int \frac{dQ_A}{T}}. \tag{1.11}$$

Parameters ξ_B and ξ_A are then defined to measure the failures to achieve the maximum and
minimum temperatures T_{max} and T_{min},

$$\xi_B \equiv \frac{\bar{T}_B}{T_{max}}, \qquad \xi_A \equiv \frac{\bar{T}_A}{T_{min}}, \tag{1.12}$$

where ξ_B is less than unity and ξ_A is greater than unity. The combined parameter

$$\xi \equiv \frac{\xi_B}{\xi_A} = \left(\frac{T_{min}}{T_{max}}\right)\left(\frac{\bar{T}_B}{\bar{T}_A}\right) = \tau\left(\frac{\bar{T}_B}{\bar{T}_A}\right), \tag{1.13}$$

where $\tau = (T_{min}/T_{max})$. ξ is then an overall measure of the failure of the real cycle to achieve the maximum and minimum temperatures and is always less than unity (except for the Carnot cycle, where ξ becomes unity).

Caputo then introduced a parameter (σ) which is a measure of the irreversibilities within the real cycle. He first defined

$$\sigma_B \equiv \frac{Q_B}{\bar{T}_B}, \qquad \sigma_A \equiv \frac{Q_A}{\bar{T}_A}, \tag{1.14}$$

which, from the definitions of $\bar{T}$, can be seen to be the entropy changes in heat supply and heat rejection, respectively. The parameter σ is then defined as

$$\sigma \equiv \frac{\sigma_B}{\sigma_A}, \tag{1.15}$$

the ratio of entropy change in heat supply to entropy change in heat rejection. For the Carnot cycle σ is unity, but for other (irreversible) cycles, a value of σ less than unity indicates a 'widening' of the cycle on the T, s diagram due to irreversibilities (e.g. in compression and/or expansion in the gas turbine cycle) and a resulting loss in thermal efficiency.

The overall effect of these failures to achieve Carnot efficiency is then encompassed in a new parameter, μ, where

$$\mu \equiv \xi\sigma. \tag{1.16}$$

The efficiency of the real cycle may then be expressed in terms of τ (the ratio of minimum to maximum temperature) and μ. For

$$\eta = 1 - \left(\frac{Q_A}{Q_B}\right) = 1 - \frac{\sigma_A \bar{T}_A}{\sigma_B \bar{T}_B} = 1 - \frac{\sigma_A \xi_A T_{min}}{\sigma_B \xi_B T_{max}} = 1 - \frac{\tau}{\sigma\xi} = 1 - \frac{\tau}{\mu}. \tag{1.17}$$

For the Carnot cycle $\sigma = 1$, $\xi = 1$ and $\mu = 1$, so that $\eta_{CAR} = 1 - \tau$.

For the JB cycle of Fig. 1.4, there is no 'widening' of the cycle due to irreversibilities, so that $\sigma_A = \sigma_B$ and $\sigma = 1$. The efficiency then is given by

$$\eta = 1 - \frac{\tau}{\xi} < \eta_{CAR} = 1 - \tau, \tag{1.18}$$

and the failure to reach Carnot cycle efficiency is entirely due to non-achievement of T_{max} and/or T_{min}. ξ is less than unity, so $\eta < \eta_{CAR}$.

For an irreversible gas turbine cycle (the irreversible Joule–Brayton (IJB) cycle of Fig. 1.9), $\sigma_A > \sigma_B$ (σ is less than unity) and $\xi < 1$ so that the thermal efficiency is

$$\eta = 1 - \frac{\tau}{\sigma\xi} = 1 - \frac{\tau}{\mu}. \tag{1.19}$$

1.5. Modifications of gas turbine cycles to achieve higher thermal efficiency

There are several modifications to the basic gas turbine cycle that may be introduced to raise thermal efficiency.

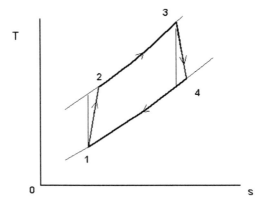

Fig. 1.9. Irreversible Joule–Brayton cycle.

Two objectives are immediately clear. If the top temperature can be raised and the bottom temperature lowered, then the ratio $\tau = (T_{min}/T_{max})$ is decreased and, as with a Carnot cycle, thermal efficiency will be increased (for given μ). The limit on top temperature is likely to be metallurgical while that on the bottom temperature is of the surrounding atmosphere.

A third objective is similarly obvious. If compression and expansion processes can attain more isentropic conditions, then the cycle 'widening' due to irreversibility is decreased, σ moves nearer to unity and the thermal efficiency increases (for a given τ). Cycle modifications or innovations are mainly aimed at increasing ξ (by increasing ξ_B or decreasing ξ_A).

Fig. 1.10 shows the processes of heat exchange (or recuperation), reheat and intercooling as additions to a JB cycle. *Heat exchange* alone, from the turbine exhaust to the compressed air before external heating, increases ξ_B and lowers ξ_A, so that the overall

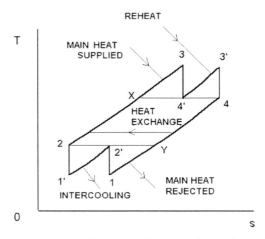

Fig. 1.10. Temperature–entropy diagram showing reheat, intercooling and recuperation.

increase in ξ leads to higher thermal efficiency. *Reheat alone* (without a heat exchanger) between two stages of turbine expansion, has the effect of increasing ξ_B but it also increases ξ_A so that ξ decreases and thermal efficiency drops. Similarly, *intercooling alone* (without a heat exchanger) lowers the mean temperature of heat rejected (decreasing ξ_A) and it also decreases ξ_B so that ξ decreases and thermal efficiency drops. However, when *reheating and intercooling are coupled with the use of a heat exchanger* then ξ_B is increased and ξ_A decreased, so ξ is increased and thermal efficiency increased markedly. Indeed, for many stages of reheat and intercooling, a Carnot cycle efficiency can in theory be attained, with all the heat supplied near the top temperature T_B and all the heat rejected near the lowest temperature, T_A.

Reheat and intercooling also increase the specific work of the cycle, the amount of work done by unit quantity of gas in passing round the plant. This is illustrated by the increase in the area enclosed by the cycle on the T, s diagram.

More details are discussed in Chapter 3, where the criteria for the performance of the components within gas turbine plants are also considered.

References

[1] Horlock, J.H. (1987), Co-generation: Combined Heat and Power, Pergamon Press, Oxford, See also 2nd edn, Krieger, Melbourne, FL, 1996.
[2] Horlock, J.H. (1992), Combined Power Plants, Pergamon Press, Oxford, See also 2nd edn, Krieger, Melbourne, FL, 2002.
[3] Haywood, R.W. (1991), Analysis of Engineering Cycles 4th edn, Pergamon Press, Oxford.
[4] Caputa, C. (1967), Una Cifra di Merito Dei Cicli Termodinamici Directti, Il Calore 7, 291–300.

Chapter 2

REVERSIBILITY AND AVAILABILITY

2.1. Introduction

In Chapter 1, the gas turbine plant was considered briefly in relation to an ideal plant based on the Carnot cycle. From the simple analysis in Section 1.4, it was explained that the closed cycle gas turbine failed to match the Carnot plant in thermal efficiency because of
(a) the 'ξ effect' (that heat is not supplied at the maximum temperature and heat is not rejected at the minimum temperature) and
(b) the 'σ effect' (related to any entropy increases within the plant, and the consequent 'widening' of the cycle on the T, s diagram).
Since these were preliminary conclusions, further explanations of these disadvantages are given using the second law of thermodynamics in this chapter. The ideas of reversibility, irreversibility, and the thermodynamic properties 'steady-flow availability' and 'exergy' are also developed.

In defining the thermal efficiency of *the closed gas turbine cycle*, such as the one shown in Fig. 1.2, we employed the first law of thermodynamics (in the form of the steady-flow energy equation round the cycle), which states that the heat supplied is equal to the work output plus the heat rejected, i.e.

$$Q_B = W + Q_A. \tag{2.1}$$

Here W is the net work output, i.e. the difference between the turbine work output (W_T) and the work required to drive the compressor (W_C), $W = W_T - W_C$.

For *the open circuit gas turbine* of Fig. 1.3, if the reactants (air M_a and fuel M_f) enter at temperature T_0, and the exhaust products ($M_a + M_f$) leave at temperature T_4, then the steady-flow energy equation yields

$$H_{R0} = W + H_{P4}, \tag{2.2}$$

where subscripts R and P refer to reactants and products, respectively, and it has been assumed that there are no heat losses from the plant. If we now consider unit air flow at entry with a fuel flow f ($= M_f/M_a$) then the enthalpy flux H_{R0} is equal to the sum of the enthalpy (h_{a0}) and the enthalpy of the fuel flow f supplied to the combustion chamber (fh_{f0}), both at ambient temperature T_0, and the enthalpy of the exhaust gas is $H_{P4} = (1+f)h_{P4}$. Hence

$$h_{a0} + fh_{f0} = w + (1+f)h_{P4}, \qquad (2.3)$$

where $w = W/M_a$ is the specific work (per unit air flow).

If the same quantities of fuel and air were supplied to a calorific value experiment at T_0 (Fig. 1.7) then the steady-flow energy equation for that process would yield

$$h_{a0} + fh_{f0} = (1+f)h_{P0} + f[CV]_0, \qquad (2.4)$$

where $[CV]_0$ is the calorific value of the fuel. Combining these two equations yields

$$f[CV]_0 = w + (1+f)(h_{P4} - h_{P0}). \qquad (2.5)$$

This equation is often used as an 'equivalent' form to Eq. (2.1), the calorific value term being regarded as the 'heat supplied' and the gas enthalpy difference term $(1+f) \times (h_{P4} - h_{P0})$ being regarded as the 'heat rejected' term.

In this chapter we will develop more rigorous approaches to the analysis of gas turbine plants using both the first and second laws of thermodynamics.

2.2. Reversibility, availability and exergy

The concepts of reversibility and irreversibility are important in the analysis of gas turbine plants. A survey of important points and concepts is given below, but the reader is referred to standard texts [1–3] for detailed presentations.

A closed system moving slowly through a series of stable states is said to undergo a reversible process if that process can be completely reversed in all thermodynamic respects, i.e. if the original state of the system itself can be recovered (internal reversibility) and its surroundings can be restored (external irreversibility). An irreversible process is one that cannot be reversed in this way.

The objective of the gas turbine designer is to make all the processes in the plant as near to reversible as possible, i.e. to reduce the irreversibilities, both internal and external, and hence to obtain higher thermal efficiency (in a closed cycle gas turbine plant) or higher overall efficiency (in an open gas turbine plant). The concepts of availability and exergy may be used to determine the location and magnitudes of the irreversibilities.

2.2.1. Flow in the presence of an environment at T_0 (not involving chemical reaction)

Consider first the steady flow of fluid through a control volume CV between prescribed stable states X and Y (Fig. 2.1) in the presence of an environment at ambient temperature T_0 (i.e. with reversible heat transfer to that environment only). The maximum work which is obtained in reversible flow between X and Y is given by

$$[(W_{CV})_{REV}]_X^Y = B_X - B_Y, \qquad (2.6)$$

where B is the steady flow availability function

$$B = H - T_0 S, \qquad (2.7)$$

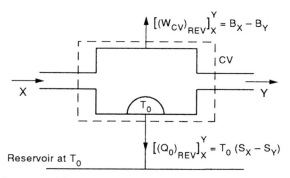

Fig. 2.1. Reversible process with heat transfer at temperature T_0 (to the environment) (after Ref. [5]).

and H and S are the enthalpy and entropy, respectively [1]. The reversible (outward) heat transfer between X and Y is

$$[(Q_0)_{REV}]_X^Y = T_0(S_X - S_Y). \tag{2.8}$$

A corollary of this theorem is that the maximum work that can be extracted from fluid at prescribed state X is the exergy

$$E_X = B_X - B_0. \tag{2.9}$$

Here B_0 is the steady flow availability function at the so-called 'dead state', where the fluid is in equilibrium with the environment, at state (p_0, T_0). The maximum work obtainable between states X and Y may then be written as

$$[(W_{CV})_{REV}]_X^Y = (B_X - B_0) - (B_Y - B_0) = (E_X - E_Y). \tag{2.10}$$

From the steady-flow energy equation, the work output in an *actual* (irreversible) flow through a control volume CV, between states X and Y in the presence of an environment at T_0 (Fig. 2.2), is

$$[W_{CV}]_X^Y = (H_X - H_Y) - [Q_0]_X^Y, \tag{2.11}$$

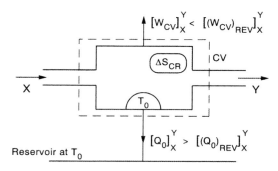

Fig. 2.2. Actual process with heat transfer at temperature T_0 (to the environment) (after Ref. [5]).

where $[Q_0]_X^Y$ is the heat transferred to the environment from the control volume. $[W_{CV}]_X^Y$ is less than $[(W_{CV})_{REV}]_X^Y$ and $[Q_0]_X^Y$ is greater than $[(Q_0)_{REV}]_X^Y$. The leaving entropy flux associated with this outward heat transfer is $[Q_0]_X^Y/T_0$, such that the increase in entropy across the control volume is

$$S_Y - S_X = \Delta S^{CR} - [Q_0]_X^Y/T_0, \tag{2.12}$$

where ΔS^{CR} is the entropy created within the control volume. The work lost due to this internal irreversibility is, therefore

$$I^{CR} = [(W_{CV})_{REV}]_X^Y - [W_{CV}]_X^Y = (B_X - B_Y) - (H_X - H_Y - [Q_0]_X^Y)$$

$$= T_0(S_Y - S_X) + [Q_0]_X^Y = [Q_0]_X^Y - [(Q_0)_{REV}]_X^Y = T_0 \Delta S^{CR}. \tag{2.13}$$

2.2.2. Flow with heat transfer at temperature T

Consider next the case where heat $[Q_{REV}]_X^Y = \int_X^Y dQ_{REV}$ is rejected (i.e. transferred *from* the control volume CV at temperature T) in a reversible steady-flow process between states X and Y, in the presence of an environment at T_0. $[Q_{REV}]_X^Y$ is taken as positive.

Fig. 2.3 shows such a fully reversible steady flow through the control volume CV. The heat transferred $[Q_{REV}]_X^Y$, supplies a reversible heat engine, delivering external work $[(W_e)_{REV}]_X^Y$ and rejecting heat $[(Q_0)_{REV}]_X^Y$ to the environment.

The total work output from the extended (dotted) control volume is $(B_X - B_Y)$, if the flow is again between states X and Y. But the work from the reversible external engine is

$$[(W_e)_{REV}]_X^Y = \int_X^Y \left(\frac{T - T_0}{T} \right) dQ_{REV}. \tag{2.14}$$

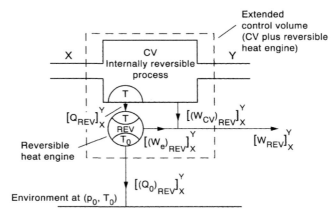

Fig. 2.3. Reversible process with heat transfer at temperature T (to Carnot engine) (after Ref. [5]).

The maximum (reversible) work obtained from the 'inner' control volume CV is therefore equal to

$$[(W_{\text{CV}})_{\text{REV}}]_X^Y = B_X - B_Y - [(W_e)_{\text{REV}}]_X^Y = E_X - E_Y - \int_X^Y \left(\frac{T - T_0}{T}\right) dQ_{\text{REV}}. \quad (2.15)$$

For a real (irreversible) flow process through the control volume CV between fluid states X and Y (Fig. 2.4), with the *same* heat rejected at temperature $T([Q]_X^Y = [Q_{\text{REV}}]_X^Y)$, the work output is $[W_{\text{CV}}]_X^Y$. Heat $[Q_0]_X^Y$ may also be transferred from CV directly to the environment at T_0. From the steady-flow energy equation,

$$[W_{\text{CV}}]_X^Y = H_X - H_Y - [Q]_X^Y - [Q_0]_X^Y. \quad (2.16)$$

The entropy flux from the control volume associated with the heat transfer is

$$\int_X^Y \frac{dQ}{T} + \frac{[Q_0]_X^Y}{T_0},$$

so the entropy increase across it is given by

$$S_Y - S_X = \Delta S^{\text{CR}} - \int_X^Y \frac{dQ}{T} - \frac{[Q_0]_X^Y}{T_0}. \quad (2.17)$$

The lost work due to irreversibility I^{CR} within the control volume CV is

$$I^{\text{CR}} = [(W_{\text{CV}})_{\text{REV}}]_X^Y - [(W)_{\text{CV}}]_X^Y$$

$$= B_X - B_Y - \int_X^Y \left(\frac{T - T_0}{T}\right) dQ - \{(H_X - H_Y) - [Q]_X^Y - [Q_0]_X^Y\}$$

$$= T_0(S_Y - S_X) + T_0 \int_X^Y \frac{dQ}{T} + [Q_0]_X^Y = T_0 \Delta S^{\text{CR}}, \quad (2.18)$$

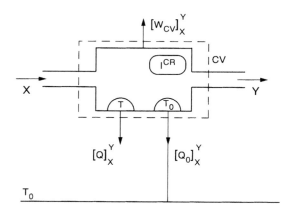

Fig. 2.4. Actual process with heat transfers at temperatures T and T_0 (after Ref. [5]).

from Eq. (2.17). Thus the work lost due to internal irreversibility within the control volume when heat transfer takes place is still $T_0 \Delta S^{CR}$, as when the heat transfer is limited to exchange with the environment.

The actual work output in a *real* irreversible process between stable states X and Y is therefore

$$[W_{CV}]_X^Y = [(W_{CV})_{REV}]_X^Y - \int_X^Y \left(\frac{T - T_0}{T} \right) dQ - I^{CR}$$

$$= B_X - B_Y - \int_X^Y \left(\frac{T - T_0}{T} \right) dQ - I^{CR} = E_X - E_Y - E_{OUT}^Q - I^{CR}, \quad (2.19)$$

where

$$E_{OUT}^Q = \int_X^Y \left(\frac{T - T_0}{T} \right) dQ$$

is the work potential, sometimes called the thermal energy of the heat rejected.

The above analysis has been concerned with heat transfer *from* the control volume. Consider next heat $[dQ]_X^Y = [dQ_{REV}]_X^Y$ transferred *to* the control volume. Then that heat could be reversibly pumped to CV (at temperature T) from the atmosphere (at temperature T_0) by a reversed Carnot engine. This would require work *input*

$$[(W_e)_{REV}]_X^Y = \int_X^Y \left(\frac{T - T_0}{T} \right) dQ_{REV},$$

Under this new arrangement, Eq. (2.15) for the reversible work delivered from CV would become

$$[(W_{CV})_{REV}]_X^Y = (E_X - E_Y) + \int_X^Y \left(\frac{T - T_0}{T} \right) dQ_{REV}, \quad (2.20)$$

and Eq. (2.19) for the work output from the actual process would be

$$[W_{CV}]_X^Y = (E_X - E_Y) + E_{IN}^Q - I^{CR}, \quad (2.21)$$

where E_{IN}^Q is the work potential or thermal energy of the heat supplied to CV,

$$E_{IN}^Q = \int_X^Y \left(\frac{T - T_0}{T} \right) dQ.$$

If heat were both transferred to and rejected from CV, then a combination of Eqs. (2.19) and (2.21) would give

$$[W_{CV}]_X^Y = (E_X - E_Y) + E_{IN}^Q - E_{OUT}^Q - I^{CR}. \quad (2.22)$$

2.3. Exergy flux

Eq. (2.22) may be interpreted in terms of exergy flows, work output and work potential (Fig. 2.5). The equation may be rewritten as

$$E_X = [W_{CV}]_X^Y + (E_{OUT}^Q - E_{IN}^Q) + I^{CR} + E_Y. \tag{2.23}$$

Thus, the exergy E_X of the entering flow (its capacity for producing work) is translated into

(i)   the actual work output $[W_{CV}]_X^Y$,
(ii)   the work potential, or thermal exergy, of the heat rejected less than that of the heat supplied $(E_{OUT}^Q - E_{IN}^Q)$,
(iii) the work lost due to internal irreversibility, $I^{CR} = T_0 \Delta S^{CR}$,
(iv) the leaving exergy, E_Y.

If the heat transferred from the control volume is not used externally to create work, but is simply lost to the atmosphere in which further entropy is created, then E_{OUT}^Q can be said to be equal to I_{OUT}^Q, a lost work term, due to external irreversibility. Another form of Eq. (2.23) is thus

$$E_X + E_{IN}^Q = W_{CV} + \sum I^{CR} + I_{OUT}^Q + E_Y, \tag{2.24}$$

which illustrates how the total exergy supplied is used or wasted.

These equations for energy flux are frequently used to trace exergy through a power plant, by finding the difference between the exergy at entry to a component $[E_X]$ and that at exit $[E_Y]$, and summing such differences for all the components to obtain an exergy statement for the whole plant, as in Sections 2.3.1 and 2.6 below. Practical examples of the application of this technique to real gas turbine plants are given below and in the later chapters.

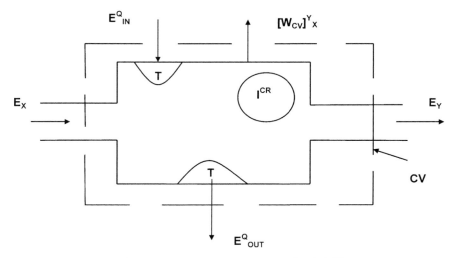

Fig. 2.5. Exergy fluxes in actual process (after Ref. [5]).

2.3.1. *Application of the exergy flux equation to a closed cycle*

We next consider the application of the exergy flux equation to a closed cycle plant based on the Joule–Brayton (JB) cycle (see Fig. 1.4), but with irreversible compression and expansion processes—an 'irreversible Joule–Brayton' (IJB) cycle. The T, s diagram is as shown in Fig. 2.6.

If the exergy flux (Eq. (2.23)) is applied to the four processes 1-2, 2-3, 3-4, 4-1, then

$$E_1 - E_2 = I_{12}^{\text{CR}} + W_{12}, \qquad E_2 - E_3 = -E_{\text{IN}}^Q, \qquad E_3 - E_4 = I_{34}^{\text{CR}} + W_{34},$$
$$E_4 - E_1 = E_{\text{OUT}}^Q. \tag{2.25}$$

Hence, by addition the exergy equation for the whole cycle is

$$E_{\text{IN}}^Q - E_{\text{OUT}}^Q = W_{\text{IJB}} + \sum I^{\text{CR}}, \tag{2.26}$$

where $W_{\text{IJB}} = W_{34} + W_{12} = W_{\text{T}} - W_{\text{C}}$, the difference between the turbine work *output* $W_{\text{T}} = W_{34}$ and the compressor work *input*, $W_{\text{C}} = -W_{12}$.

The corresponding 'first law' equations for the closed cycle gas turbine plant lead to

$$Q_{\text{IN}} - Q_{\text{OUT}} = W_{\text{T}} - W_{\text{C}} = W_{\text{IJB}}, \tag{2.27}$$

in comparison with Eq. (2.26).

2.3.2. *The relationships between ξ, σ and I^{CR}, I^Q*

The exergy equation (2.26) enables useful information on the irreversibilities and lost work to be obtained, in comparison with a Carnot cycle operating within the same temperature limits ($T_{\text{max}} = T_3$ and $T_{\text{min}} = T_0$). Note first that if the heat supplied Q_B is the same to each of the two cycles (Carnot and IJB), then the work output from the Carnot engine (W_{CAR}) is greater than that of the IJB cycle (W_{IJB}), and the heat rejected from the former is less than that rejected by the latter.

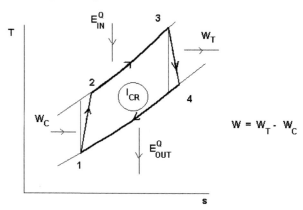

Fig. 2.6. Exergy fluxes in closed IJB gas turbine cycle.

An exergy flux statement for the Carnot plant is

$$[E_{IN}^Q]_{CAR} = W_{CAR}, \tag{2.28}$$

where $[E_{IN}^Q]_{CAR} = \int(1 - (T_0/T_3))dQ = \eta_{CAR}Q_B$ and E_{OUT}^Q is zero.
For the IJB cycle

$$[E_{IN}^Q]_{IJB} = W_{IJB} + [E_{OUT}^Q]_{IJB} + \left[\sum I^{CR}\right]_{IJB}. \tag{2.29}$$

The difference between Eqs. (2.28) and (2.29) is

$$[E_{IN}^Q]_{CAR} - [E_{IN}^Q]_{IJB} = W_{CAR} - W_{IJB} - \left[E_{OUT}^Q + \sum I^{CR}\right]_{IJB}.$$

Hence,

$$W_{IJB} = W_{CAR} - I_{IN}^Q - I_{OUT}^Q - \sum I^{CR}, \tag{2.30}$$

where

$$I_{IN}^Q = [E_{IN}^Q]_{CAR} - [E_{IN}^Q]_{IJB} = \int\left(\frac{T_0}{T} - \frac{T_0}{T_3}\right)dQ_B, \tag{2.31}$$

and

$$I_{OUT}^Q = [E_{OUT}^Q]_{IJB} = \int\left(1 - \frac{T_0}{T}\right)dQ_A. \tag{2.32}$$

I_{IN}^Q, I_{OUT}^Q may be regarded as irreversibilities of heat supply and rejection in the IJB cycle. I_{IN}^Q is the lost work involved in supplying heat Q_B from a reservoir at a constant (maximum) temperature T_3 to the IJB air heater at temperature T, rather than to a Carnot cycle air heater at a temperature just below T_3. I_{OUT}^Q is the lost work involved in rejection of the (larger) quantity of heat Q_A from the IJB cycle to the atmosphere.

The thermal efficiency of the IJB cycle is thus less than that of the Carnot plant, by an amount

$$\eta_{CAR} - \eta_{IJB} = \frac{W_{CAR}}{Q_B} - \frac{W_{IJB}}{Q_B} = \frac{(I_{IN}^Q + I_{OUT}^Q)}{Q_B} + \frac{\sum I^{CR}}{Q_B} \tag{2.33a}$$

$$= (\tau/\xi\sigma)_{IJB} - \tau, \tag{2.33b}$$

where ξ and σ are the parameters that were introduced in the simple preliminary analysis of the IJB cycle given in Chapter 1, Section 1.4. ξ was related to the mean temperatures of supply and rejection and σ to the 'widening' of the cycle.

Thus for a JB cycle, with no internal irreversibility, $I^{CR} = 0$ and $\sigma_{JB} = 1$, from Eqs. (2.33) and (1.17)

$$\eta_{CAR} - \eta_{JB} = \frac{(I_{IN}^Q + I_{OUT}^Q)}{Q_B} = \tau\left[\frac{\xi_A}{\xi_B} - 1\right] = \tau[(1/\xi_{IJB}) - 1]. \tag{2.34}$$

For an 'irreversible' Carnot type cycle (ICAR) with all heat supplied at the top temperature and all heat rejected at the lowest temperature ($T_{max} = T_3$, $T_{min} = T_0$, $I^Q_{OUT} = 0$, $\xi_{ICAR} = 1$), but with irreversible compression and expansion ($\sigma_{ICAR} = \sigma_B/\sigma_A < 1$), Eqs. (2.33) and (1.17) yield

$$\eta_{CAR} - \eta_{ICAR} = \frac{\sum I^{CR}}{Q_A} = \tau\left[\left(\frac{\sigma_A}{\sigma_B}\right) - 1\right] = \tau[(1/\sigma_{ICAR}) - 1]. \tag{2.35}$$

However, use of Eqs. (2.34) and (2.35) together does not yield Eq. (2.33b) because the values of I^Q and $\sum I^{CR}$ are not the same in the IJB, JB and ICAR cycles.

2.4. The maximum work output in a chemical reaction at T_0

The (maximum) reversible work in steady flow between reactants at an entry state $R_0(p_0, T_0)$ and products at a leaving state $P_0(p_0, T_0)$ is

$$[(W_{CV})_{REV}]^{P0}_{R0} = B_{R0} - B_{P0}. \tag{2.36}$$

It is supposed here that the various reactants entering are separated at (p_0, T_0); the various products discharged are similarly separated at (p_0, T_0). The maximum work may then be written as

$$B_{R0} - B_{P0} = G_{R0} - G_{P0} = [-\Delta G_0], \tag{2.37}$$

where G is the Gibbs function, $G = H - TS$. This is the maximum work obtainable from such a combustion process and is usually used in defining the rational efficiency of an open circuit plant. However, it should be noted that if the reactants and/or products are not at pressure p_0, then the work of delivery or extraction has to be allowed for in obtaining the maximum possible work from the reactants and products drawn from and delivered to the atmosphere. The expression for maximum work has to be modified.

Kotas [3] has drawn a distinction between the 'environmental' state, called the dead state by Haywood [1], in which reactants and products (each at p_0, T_0) are in *restricted* thermal and mechanical equilibrium with the environment; and the 'truly or completely dead state', in which they are also in chemical equilibrium, with partial pressures (p_k) the same as those of the atmosphere. Kotas defines the chemical exergy as the sum of the maximum work obtained from the reaction with components at p_0, T_0, $[-\Delta G_0]$, and work extraction and delivery terms. The delivery work term is $\sum_k M_k R_k T_0 \ln(p_0/p_k)$, where p_k is a partial pressure, and is positive. The extraction work is also $\sum_k M_k R_k T_0 \ln(p_0/p_k)$ but is negative.

In general, we shall not subsequently consider these extraction and delivery work terms here, but use $[-\Delta G_0]$ as an approximation to the maximum work output obtainable from a chemical reaction, since the work extraction and delivery quantities are usually small. Their relative importance is discussed in detail by Horlock et al. [4].

2.5. The adiabatic combustion process

Returning to the general availability equation, for an adiabatic combustion process between reactants at state X and products at state Y (Fig. 2.7) we may write

$$B_{RX} - B_{PY} = I^{CR} = T_0 \Delta S^{CR},$$ (2.38)

since there is no heat or work transfer, and the work lost due to internal irreversibility is I^{CR}. In forming the exergy at the stations X and Y we must be careful to subtract the steady flow availability function in the final equilibrium state, which we take here as the product (environmental) state at (p_0, T_0). Then Eq. (2.38) may be written as

$$B_{RX} - G_{P0} = B_{PY} - G_{P0} + T_0 \Delta S^{CR}$$

or

$$B_{RX} - B_{R0} + [-\Delta G_0] = B_{PY} - B_{P0} + T_0 \Delta S^{CR}.$$ (2.39)

It is convenient for exergy tabulations to associate the term $[-\Delta G_0] = G_{R0} - G_{P0}$ with the exergy of the fuel supplied (of mass M_f), i.e. $E_{f0} = [-\Delta G_0]$. For a combustion process burning liquid or solid fuel (at temperature T_0) with air (subscript a, at temperature T_1), the left-hand side of the equation may be written as

$$E_X = B_{a1} - B_{a0} + [-\Delta G_0] = E_{a1} + E_{f0}.$$ (2.40)

Usually, $T_1 = T_0$, so $E_{a1} = E_{a0}$ and with $E_Y = E_{P2}$ the exergy equation becomes

$$E_{a0} + E_{f0} = E_{P2} + I^{CR}.$$ (2.41)

For a combustion process burning gaseous fuel (which may have been compressed from state 0 to state $1'$), the left-hand side of the exergy Eq. (2.41) may be rewritten as

$$E_{a0} + E_{f1'} = E_{P2} + I^{CR}.$$ (2.42)

In general, for any gas of mass M we may write

$$E = M(h - h_0) - MT_0(s - s_0),$$ (2.43)

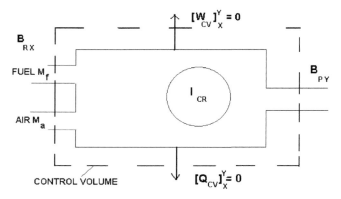

Fig. 2.7. Exergy fluxes in adiabatic combustion.

where h_0 and s_0 are the specific enthalpy and specific entropy at the ambient pressure p_0 and the temperature T_0, respectively. For a semi-perfect gas with $p = \rho RT$ and $c_p = c_p(T)$,

$$h - h_0 = \int_{T_0}^{T} c_p dT, \tag{2.44}$$

$$s - s_0 = \phi - R \ln(p/p_0), \tag{2.45}$$

$$b - b_0 = \int_{T_0}^{T} c_p dT - T_0 \phi + RT_0 \ln(p/p_0), \tag{2.46}$$

where

$$\phi \equiv \int_{T_0}^{T} \frac{c_p dT}{T}.$$

2.6. The work output and rational efficiency of an open circuit gas turbine

The statements on work output made for a real process (Eq. (2.23)) and for the ideal chemical reaction or combustion process at (p_0, T_0) (Eq. (2.37)) can be compared as

$$B_{RX} = B_{PY} + [W_{CV}]_X^Y + \int_X^Y \frac{T - T_0}{T} dQ + T_0 \sum \Delta S^{CR}; \tag{2.47}$$

$$G_{R0} = G_{P0} + [(W_{CV})_{REV}]_{R0}^{P0}. \tag{2.48}$$

The first equation may be applied to a control volume CV surrounding a gas turbine power plant, receiving reactants at state $R_X \equiv R_0$ and discharging products at state $P_Y \equiv P_4$. As for the combustion process, we may subtract the steady flow availability function for the equilibrium product state (G_{P0}) from each side of Eq. (2.47) to give

$$[-\Delta G_0] = G_{R0} - G_{P0} = [(W_{CV})_{REV}]_{R0}^{P0}$$

$$= [W_{CV}]_X^Y + E_{OUT}^Q + T_0 \Delta S^{CR} + [B_{P4} - G_{P0}]. \tag{2.49}$$

This equation, as illustrated in the (T, s) chart of Fig. 2.8 for an open circuit gas turbine, shows how the maximum possible work output from the ideal combustion process splits into the various terms on the right-hand side:
- the actual work output from the open circuit gas turbine plant;
- the work potential of any heat transferred out from various components, which if transferred to the atmosphere at T_0, becomes the work lost due to external irreversibility, $E_{OUT}^Q = I_{OUT}^Q$;
- the work lost due to internal irreversibility, I^{CR} (which may occur in various components);
- the work potential of the discharged exhaust gases, $(B_{P4} - G_{P0})$.

Note that in Eq. (2.49) the term $(B_{RX} - G_{R0})$ does not appear as it has been assumed here that all reactants enter at the ambient temperature T_0, for which $[-\Delta G_0]$ is known. For a compressed gaseous fuel, $(B_{RX} - G_{R0})$ will be small but not entirely negligible.

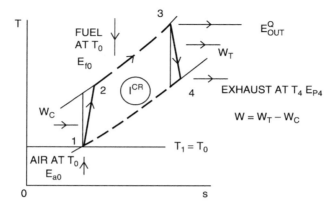

Fig. 2.8. Exergy fluxes in actual CBT gas turbine plant with combustion.

The rational efficiency may be defined as the ratio of the actual work output $[W_{CV}]_X^Y$ to the maximum possible work output, approximately $[-\Delta G_0]$,

$$(\eta_R) \approx \frac{[W_{CV}]_X^Y}{[-\Delta G_0]} \approx 1 - \frac{I^Q}{[-\Delta G_0]} - \frac{\sum I^{CR}}{[-\Delta G_0]} - \frac{(B_P)_4 - (G_P)_0}{[-\Delta G_0]}. \qquad (2.50)$$

Fig. 2.9 illustrates this approach of tracing exergy through a plant. The various terms in Eq. (2.49) are shown for an irreversible open gas turbine plant based on the JB cycle. The compressor pressure ratio is 12:1, the ratio of maximum to inlet temperature is 5:1 ($T_{max} = 1450$ K with $T_0 = 290$ K), the compressor and turbine polytropic efficiencies are

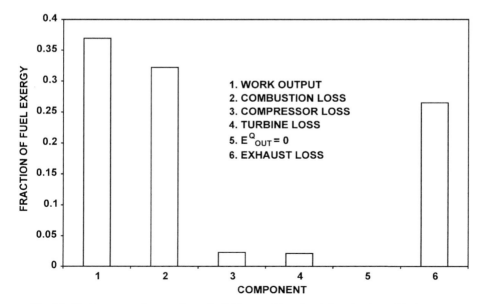

Fig. 2.9. Work output and exergy losses in CBT gas turbine plant (all as fractions of fuel exergy).

0.9, and the combustion pressure loss is 3% of the inlet pressure to the chamber. The method of calculation is given in Chapters 4 and 5, but it is sufficient to say here that it involves the assumption of real semi-perfect gases with methane as fuel for combustion and no allowance for any turbine cooling. The work terms associated with the abstraction and delivery to the atmosphere are ignored in the valuation of the fuel exergy, which is thus taken as $[-\Delta G_0]$.

The thermal efficiency, the work output as a fraction of the fuel exergy (the maximum reversible work), is shown as no. 1 in the figure and is 0.368. The internal irreversibility terms, $\sum I^{CR}/[-\Delta G_0]$, are shown as nos. 2, 3, and 4 in the diagram, for the combustion chamber, compressor and turbine, respectively. It is assumed that there is no *heat* rejection to the atmosphere from the engine, i.e. $I^Q = 0$ (no. 5), but there is an exergy loss in the discharge of the exhaust gas to the atmosphere, $(B_{P4} - G_{P0})/[-\Delta G_0]$, the last term of Eq. (2.49), which is shown as no. 6 in the diagram.

The dominant irreversibilities are in combustion and in the exhaust discharge.

2.7. A final comment on the use of exergy

We shall later give more detailed calculations for real gas turbine plants together with diagrams similar to Fig. 2.9. Exergy is a very useful tool in determining the magnitude of local losses in gas turbine plants, and in his search for high efficiency the gas turbine designer seeks to reduce these irreversibilities in components (e.g. compressor, turbine, the combustion process, etc.).

However, it is wise to emphasise the interactions between such components. An improvement in one (say an increase in the effectiveness of the heat exchanger in a [CBTX]$_I$ recuperative plant) will lead to a local reduction in the irreversibility or exergy loss within it. But this will also have implications elsewhere in the plant. For the [CBTX]$_I$ plant, an increase in the recuperator effectiveness will lead to a higher temperature entering the combustion chamber and a lower temperature of the gas leaving the hot side of the exchanger. The irreversibility in combustion is decreased and the exergy loss in the final exhaust gas discharged to atmosphere is also reduced [6].

Therefore, plots of exergy loss or irreversibility like Fig. 2.9, for a particular plant operating condition, do not always provide the complete picture of gas turbine performance.

References

[1] Haywood, R.W. (1980), Equilibrium Thermodynamics, Wiley, New York.
[2] Gyftopoulos, E.P. and Beretta, G.P. (1991), Thermodynamic Foundations and Applications, MacMillan, New York.
[3] Kotas, T.J. (1985), The Exergy Method of Thermal Power Analysis, Butterworth, London.
[4] Horlock, J.H., Manfrida, G. and Young, J.B. (2000), Exergy analysis of modern fossil-fuel power plants, ASME J. Engng Gas Turbines Power 122, 1–17.
[5] Horlock, J.H. (2002), Combined Power Plants, 2nd edn, Krieger, Melbourne USA.
[6] Horlock, J.H. (1998), The relationship between effectiveness and exergy loss in counterflow heat exchangers, ASME Paper 1998-GT-32.

Chapter 3

BASIC GAS TURBINE CYCLES

3.1. Introduction

In the introduction to Chapter 1 on power plant thermodynamics our search for high thermal efficiency led us to emphasis on raising the maximum temperature T_{max} and lowering the minimum temperature T_{min}, in emulation of the performance of the Carnot cycle, the efficiency of which increases with the ratio (T_{max}/T_{min}). In a gas turbine plant, this search for high maximum temperatures is limited by material considerations and cooling of the turbine is required. This is usually achieved in 'open' cooling systems, using some compressor air to cool the turbine blades and then mixing it with the mainstream flow.

Initially in this chapter, analyses of basic gas turbine cycles are presented by reference to closed uncooled 'air standard' (a/s) cycles using a perfect gas (one with both the gas constant R and the specific heats c_p and c_v constant) as the working fluid in an externally heated plant. Many of the broad conclusions reached in this way remain reasonably valid for an open cycle with combustion, i.e. for one involving real gases with variable composition and specific heats varying with temperature. The a/s arguments are developed sequentially, starting with reversible cycles in Section 3.2 and then introducing irreversibilities in Section 3.3.

In Section 3.4, we consider the open gas turbine cycle in which fuel is supplied in a combustion chamber and the working fluids before and after combustion are assumed to be separate semi-perfect gases, each with $c_p(T)$, $c_v(T)$, but with $R = [c_p(T) - c_v(T)]$ constant. Some analytical work is presented, but recently the major emphasis has been on computer solutions using gas property tables; results of such computations are presented in Section 3.5.

Subsequently, in Chapter 4, we deal with cycles in which the turbines are cooled. The basic thermodynamics of turbine cooling, and its effect on plant efficiency, are considered. In Chapter 5, some detailed calculations of the performance of gas turbines with cooling are presented.

We adopt the nomenclature introduced by Hawthorne and Davis [1], in which compressor, heater, turbine and heat exchanger are denoted by C, H, T and X, respectively, and subscripts R and I indicate internally reversible and irreversible processes. For the open cycle, the heater is replaced by a burner, B. Thus, for example, [CBTX]₁ indicates an open irreversible regenerative cycle. Later in this book, we shall in addition, use subscripts

27

U and C referring to uncooled and cooled turbines in a plant, but in this chapter, all cycles are assumed to be uncooled and these subscripts are not used.

It is implied that the states referred to in any cycle are stagnation states; but as velocities are assumed to be low, stagnation and static states are virtually identical.

3.2. Air standard cycles (uncooled)

3.2.1. Reversible cycles

3.2.1.1. The reversible simple (Joule–Brayton) cycle, [CHT]$_R$

We use the original Joule–Brayton cycle as a standard—an internally reversible closed gas turbine cycle 1, 2, 3, 4 (see the T, s diagram of Fig. 3.1), with a maximum temperature $T_3 = T_B$ and a pressure ratio r. The minimum temperature is taken as T_A, the ambient temperature, so that $T_1 = T_A$.

For unit air flow rate round the cycle, the heat supplied is $q_B = c_p(T_3 - T_2)$, the turbine work output is $w_T = c_p(T_3 - T_4)$ and the compressor work input is $w_C = c_p(T_2 - T_1)$. Hence the thermal efficiency is

$$\eta = w/q_B = [c_p(T_3 - T_4) - c_p(T_2 - T_1)]/[c_p(T_3 - T_2)]$$

$$= 1 - [(T_4 - T_1)/(T_3 - T_2)] = 1 - \{[(T_3/xT_1) - 1]/[(T_3/T_1) - x]\} = (1/x), \quad (3.1)$$

where $x = r^{(\gamma-1)/\gamma} = T_2/T_1 = T_3/T_4$ is the isentropic temperature ratio.

Initially this appears to be an odd result as the thermal efficiency is independent of the maximum and minimum temperatures. However, each elementary part of the cycle, as shown in the figure, has the same ratio of temperature of supply to temperature of rejection

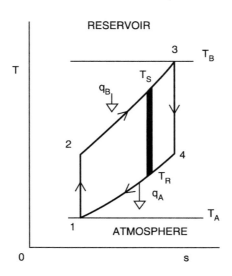

Fig. 3.1. T, s diagram for reversible closed simple cycle, [CHT]$_R$.

($T_S/T_R = x$). Thus each of these elementary cycles has the same Carnot type efficiency, equal to $[1 - (T_R/T_S)] = [1 - (1/x)]$. Hence it is not surprising that the whole reversible cycle, made up of these elementary cycles of identical efficiency, has the same efficiency. However, the net specific work,

$$w = (w_T - w_C) = c_p T_1 [(\theta/x) - 1](x - 1), \tag{3.2}$$

does increase with $\theta = T_3/T_1$ at a given x. For a given θ, it is a maximum at $x = \theta^{1/2}$.

Although the $[CHT]_R$ cycle is internally reversible, *external* irreversibility is involved in the heat supply from the external reservoir at temperature T_B and the heat rejection to a reservoir at temperature T_A. So a consideration of the *internal* thermal efficiency alone does not provide a full discussion of the thermodynamic performance of the plant. If the reservoirs for heat supply and rejection are of infinite capacity, then it may be shown that the irreversibilities in the heat supply (q_B) and the heat rejection (q_A), respectively, both positive, are

$$i_B = T_A \int (dq_B/T) - q_B T_A/T_B = [c_p T_A \ln(T_B/T_2) - (q_B T_A/T_B)], \tag{3.3}$$

and

$$i_A = q_A - T_A \int (dq_A/T) = q_A - c_p T_A \ln(T_4/T_A). \tag{3.4}$$

But since $T_B = T_3 = xT_4$ and $T_2 = xT_1 = xT_A$, so that $(T_B/T_2) = (T_4/T_A)$, it follows that

$$\sum i = i_B + i_A = q_B[(q_A/q_B) - (T_A/T_B)]. \tag{3.5}$$

With the two reservoirs at T_A and T_B, the maximum possible work is then

$$w_{max} = w + \sum i = q_B - q_A + q_A - q_B T_A/T_B = q_B[1 - (T_A/T_B)] = \eta_{CAR} q_B, \tag{3.6}$$

where η_{CAR} is the Carnot efficiency,

$$\eta_{CAR} = [1 - (T_A/T_B)]. \tag{3.7}$$

As a numerical example, for a reversible cycle with $T_B = 1800$ K, $T_A = 300$ K and $x = 2.79$ ($r = 36.27$), it follows that

$$\eta = 0.642,$$

$$i_B/q_B = 0.072, \qquad i_A/q_B = 0.120, \qquad \sum i/q_B = 0.192,$$

$$\eta_{CAR} = (\eta)_{RU} + \sum i/q_B = 0.833.$$

3.2.1.2. *The reversible recuperative cycle [CHTX]_R*

A reversible recuperative a/s cycle, with the maximum possible heat transfer from the exhaust gas, $q_T = c_p(T_4 - T_Y)$, is illustrated in the T, s diagram of Fig. 3.2, where $T_Y = T_2$. This heat is transferred to the compressor delivery air, raising its temperature to $T_X = T_4$, before entering the heater. The net specific work output is the same as that

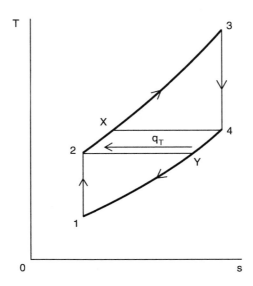

Fig. 3.2. T,s diagram for reversible closed recuperative cycle, [CHTX]$_R$.

of the [CHT]$_R$ cycle (the area enclosed on the T,s diagram is the same) but the heat supplied from the external reservoir to reach the temperature $T_3 = T_A$ is now less than in the [CHT]$_R$ cycle. It is apparent from the T,s diagram that the heat supplied, $q_B = c_p(T_3 - T_X)$ is equal to the turbine work output, $w_T = c_p(T_3 - T_4)$, and hence the thermal efficiency is

$$\eta = (w_T - w_C)/w_T = 1 - (w_C/w_T) = 1 - \{T_1(x - 1)/T_3[1 - (1/x)]\}$$

$$= 1 - (x/\theta). \tag{3.8}$$

The internal thermal efficiency increases as θ is increased, but unlike the [CHT]$_R$ cycle efficiency, drops with increase in pressure ratio r. This is because the heat transferred q_T decreases as r is increased.

Plots of thermal efficiency for the [CHT]$_R$ and [CHTX]$_R$ cycles against the isentropic temperature ratio x are shown in Fig. 3.3, for $\theta = T_3/T_1 = 4, 6.25$. The efficiency of the [CHT]$_R$ cycle increases continuously with x independent of θ, but that of the [CHTX]$_R$ cycle increases with θ for a given x. For a given $\theta = T_3/T_1$, the efficiency of the [CHTX]$_R$ cycle is equal to the Carnot efficiency at $x = 1$ and then decreases with x until it meets the efficiency line of the [CHT]$_R$ at $x = (\theta)^{1/2}$ where $\eta = 1 - (1/\theta)^{1/2}$. When $x > \theta^{1/2}$, where $T_4 = T_2$, a heat exchanger cannot be used.

The specific work of the two cycles is the same (Eq. (3.2)), and reaches a maximum at $x = (\theta)^{1/2}$ where $(w/c_pT_1) = (\theta^{1/2} - 1)^2$.

3.2.1.3. The reversible reheat cycle [CHTHT]$_R$

If reheat is introduced between a high pressure turbine and a low pressure turbine then examination of the T,s diagram (Fig. 3.4a) shows that the complete cycle is now made up

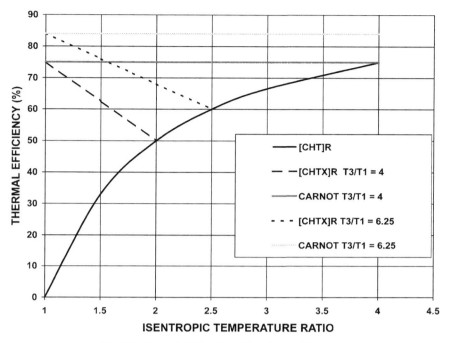

Fig. 3.3. Thermal efficiencies of closed reversible cycles.

of two types of elementary cycles, $1, 2, 3, 4''$ and $4'', 4', 3', 4$. The efficiency of the latter cycle is $\eta = 1 - (1/x_A)$, where $x_A = T_{4'}/T_{4''} = T_{3'}/T_4$; it is less than that of the former cycle $\eta = 1 - (1/x)$ and the overall efficiency of the 'combined' cycle is therefore

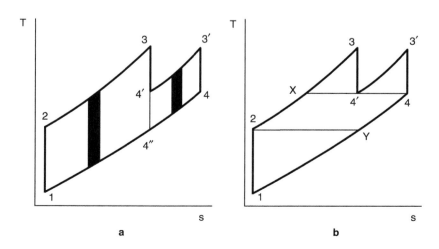

Fig. 3.4. T, s diagram for reheating added to reversible simple and recuperative cycles.

reduced compared with that of the [CHT]$_R$ cycle. However, the specific work, which is equal to the area of the cycle on the T, s diagram, is increased.

If a heat exchanger is added at low pressure ratio (Fig. 3.4b) then the mean supply temperature is greater than that of the [CHTX]$_R$ cycle whereas the temperature of heat rejection will be the same as in the [CHTX]$_R$ cycle. Therefore the efficiency of the [CHTHTX]$_R$ cycle is greater than that of the [CHTX]$_R$ cycle.

3.2.1.4. The reversible intercooled cycle [CICHT]$_R$

If the compression is split and intercooling is introduced between a low pressure compressor and a high pressure compressor (Fig. 3.5a), then by considering the elementary cycles it can be seen that the efficiency should be reduced compared with the [CHT]$_R$ cycle.

However, addition of a heat exchanger at low pressure ratio (Fig. 3.5b) means that while the mean temperature of heat supply remains the same as in the [CHTX]$_R$ cycle, the temperature of heat rejection is lowered compared with that cycle. The efficiency of the [CICHTX]$_R$ cycle is therefore greater than that of the [CHTX]$_R$ cycle.

3.2.1.5. The 'ultimate' gas turbine cycle

In the 'ultimate' version of the reheated and intercooled reversible cycle [CICICIC$\cdots$HTHTHT$\cdots$X]$_R$, both the compression and expansion are divided into a large number of small processes, and a heat exchanger is also used (Fig. 3.6). Then the efficiency approaches that of a Carnot cycle since all the heat is supplied at the maximum temperature $T_B = T_{max}$ and all the heat is rejected at the minimum temperature $T_A = T_{min}$.

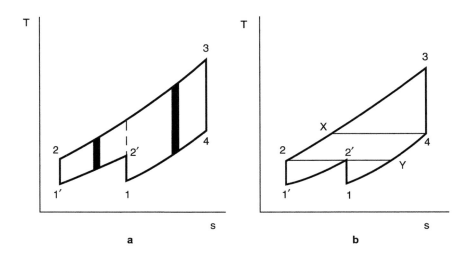

Fig. 3.5. T, s diagram for intercooling added to reversible simple and recuperative cycles.

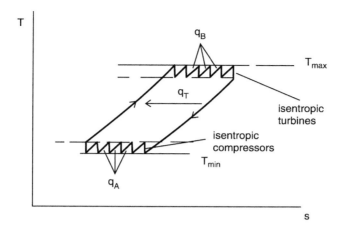

Fig. 3.6. T, s diagram for 'ultimate' reversible gas turbine cycle [CICIC···BTBT···X]$_R$.

3.2.2. *Irreversible air standard cycles*

3.2.2.1. *Component performance*

Before moving on to the a/s analyses of irreversible gas turbine cycles we need to define various criteria for the performance of some components, all of which have been assumed to be perfect (reversible) in the analyses of Section 3.2.1. The criteria used are listed in Table 3.1.

In addition to the irreversibilities associated with these components, pressure losses (Δp) may occur in various parts of the plant (e.g. in the entry and exit ducting, the combustion chamber, and the heat exchanger). These are usually expressed in terms of non-dimensional pressure loss coefficients, $\xi = \Delta p/(p)_{IN}$, where $(p)_{IN}$ is the pressure at entry to the duct. (Mach numbers are assumed to be low, with static and stagnation pressures and their loss coefficients approximately the same.)

As alternatives to the isentropic efficiencies for the turbomachinery components, η_T and η_C, which relate the overall enthalpy changes, small-stage or polytropic efficiencies (η_{pT} and η_{pC}) are often used. The pressure–temperature relationship along an expansion line is then

$$p/T^z = \text{constant}, \quad \text{where } z = [\gamma/(\gamma - 1)\eta_{pT}],$$

and the entry and exit temperatures are related by $T_3/T_4 = r_T^{(1/z)} = x_T$.

Table 3.1
Performance criteria

Component	Criterion of performance
Turbine	Isentropic efficiency η_T = Enthalpy drop/isentropic enthalpy drop
Compressor	Isentropic efficiency η_C = Isentropic enthalpy rise/enthalpy rise
Heat exchanger	Effectiveness (or thermal ratio) ε = Temperature rise (cold side)/maximum temperature difference between entry (hot side) and entry (cold side)

Along a compression line,

$$p/T^z = \text{constant}, \text{ where now } z = [\gamma\eta_{pC}/(\gamma - 1)],$$

and exit and entry temperatures are related by $T_2/T_1 = r_C^{(1/z)} = x_C$.

The analysis of Hawthorne and Davis [1] for irreversible a/s cycles is developed using the criteria of component irreversibility, firstly for the simple cycle and subsequently for the recuperative cycle. In the main analyses, the isentropic efficiencies are used for the turbomachinery components. Following certain significant relationships, alternative expressions, involving polytropic efficiency and x_C and x_T, are given, without a detailed derivation, in equations with **p** added to the number.

3.2.2.2. *The irreversible simple cycle [CHT]$_I$*

A closed cycle [CHT]$_I$, with state points $1, 2, 3, 4$, is shown in the T, s diagram of Fig. 3.7. The specific compressor work input is given by

$$w_C = c_p(T_2 - T_1) = c_p(T_{2s} - T_1)/\eta_C = c_pT_1(x - 1)/\eta_C. \tag{3.9}$$

The specific turbine work output is

$$w_T = c_p(T_3 - T_4) = \eta_T c_p(T_3 - T_{4s}) = \eta_T c_p T_3[1 - (1/x)], \tag{3.10}$$

so that the net specific work is

$$w = w_T - w_C = c_pT_1\{\eta_T\theta[1 - (1/x)] - [(x - 1)/\eta_C]\}$$

$$= c_pT_1\{\alpha[1 - (1/x)] - (x - 1)\}/\eta_C, \tag{3.11}$$

where $\alpha = \eta_C\eta_T\theta$, or

$$w = c_pT_1\{\theta[1 - (1/x_T)] - (x_C - 1)\}. \tag{3.11p}$$

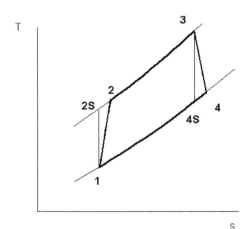

Fig. 3.7. T, s diagram for irreversible closed simple cycle [CHT]$_I$.

The specific heat supplied is

$$q = c_p(T_3 - T_2) = c_p T_1[(\theta - 1) - (x - 1)/\eta_C], \qquad (3.12)$$

or

$$q = c_p[T_3 - T_2] = c_p T_1[(\theta - 1) - (x_C - 1)], \qquad (3.12\mathbf{p})$$

so that the thermal efficiency is given by

$$\eta = w/q = (\alpha - x)[1 - (1/x)]/(\beta - x), \qquad (3.13)$$

where $\beta = 1 + \eta_C(\theta - 1)$, or

$$\eta = w/q = [\theta(1 - (1/x_T)) - (x_C - 1)]/[(\theta - 1) - (x_C - 1)]. \qquad (3.13\mathbf{p})$$

The important point here is that the efficiency is a function of the temperature ratio θ as well as the pressure ratio r (and x), whereas it is a function of pressure ratio only for the reversible cycle, $[CHT]_R$.

Optimum conditions and graphical plot

The isentropic temperature rise for maximum specific work (x_w) is obtained by differentiating Eq. (3.11) with respect to x and equating the differential to zero, giving

$$x_w = \alpha^{1/2}. \qquad (3.14)$$

By differentiating Eq. (3.13) with respect to x and equating the differential to zero, it may be shown that the isentropic temperature ratio for maximum thermal efficiency (x_e) is given by the equation

$$Ax_e^2 + Bx_e + C = 0, \qquad (3.15)$$

where $A = (\alpha - \beta - 1)$, $B = -2\alpha$, $C = \alpha\beta$.

Solution of this equation gives

$$x_e = \alpha\beta/\{\alpha + [\alpha(\beta - \alpha)(\beta - 1)]^{1/2}\}. \qquad (3.16)$$

In their graphical interpretations, using isentropic rather than polytropic efficiencies, Hawthorne and Davis plotted the following non-dimensional quantities, all against the parameter $x = r^{(\gamma-1)/\gamma}$:

Non-dimensional compressor work,

$$\text{NDCW} = w_C/c_p(T_3 - T_1) = (x - 1)/(\beta - 1); \qquad (3.17)$$

Non-dimensional turbine work,

$$\text{NDTW} = w_T/c_p(T_3 - T_1) = \alpha(x - 1)/x(\beta - 1); \qquad (3.18)$$

Non-dimensional net work,

$$\text{NDNW} = w/c_p(T_3 - T_1) = \{\alpha[1 - (1/x)] - \{x - 1\}\}/(\beta - 1); \qquad (3.19)$$

Non-dimensional heat transferred,

$$\text{NDHT} = q/c_p(T_3 - T_1) = (\beta - x)/(\beta - 1); \qquad (3.20)$$

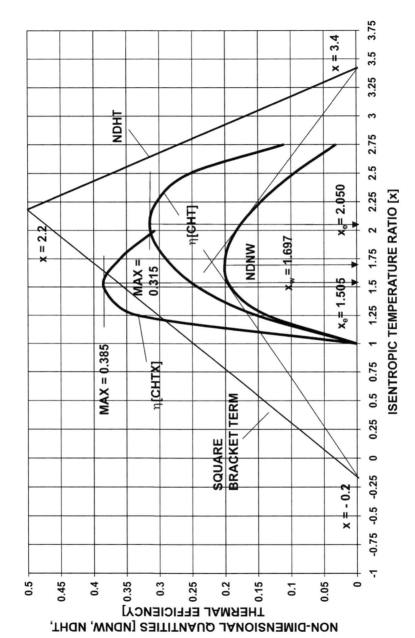

Fig. 3.8. Graphical plot for [CHT]₁ cycle (after Horlock and Woods [2]).

Thermal efficiency

$$\eta = NDNW/NDHT = [(\alpha - x)(x - 1)]/x(\beta - x). \tag{3.21}$$

Fig. 3.8 reproduces the quantities NDNW, NDHT and η, for the example of the simple $[CHT]_I$ cycle studied by Horlock and Woods [2], in which $\theta = 4.0$, $\eta_C = 0.8$, $\eta_T = 0.9$, i.e. $\alpha = 2.88$, $\beta = 3.4$. The location of the maximum net work output is obvious. The maximum cycle efficiency point is obtained by the graphical construction shown (a line drawn tangent to NDNW at x_e, from the point where the line NDHT meets the x axis at $x = \beta$). Values of $x_w = 1.697$ ($r_w = 6.368$) and $x_e = 2.050$ ($r_e = 12.344$), as calculated from Eqs. (3.14) and (3.16), are indicated in the diagrams. The maximum thermal efficiency is $\eta = 0.315$.

As mentioned before, the thermal efficiency for the irreversible plant $[CHT]_I$ is a function of the temperature ratio $\theta = T_3/T_1$ (as opposed to that of the reversible simple cycle $[CHT]_R$, for which η is a function of x only, and pressure ratio r, as illustrated in Fig. 3.3). Fig. 3.9 illustrates this difference, showing the irreversible thermal efficiency $\eta(x, \theta)$ which is strongly θ-dependent.

3.2.2.3. *The irreversible recuperative cycle [CHTX]$_I$*

For the closed recuperative cycle $[CHTX]_I$, with states $1, 2, X, 3, 4, Y$ as in the T, s diagram of Fig. 3.10, the net specific work is unchanged but the heat supplied has to be reassessed as heat q_T is transferred from the turbine exhaust to the compressor delivery air. Using the heat exchanger effectiveness, $\varepsilon = (T_X - T_2)/(T_4 - T_2)$ the heat supplied becomes

$$q_B = c_p(T_3 - T_X) = c_p(T_3 - T_4) + c_p(1 - \varepsilon)(T_4 - T_2), \tag{3.22}$$

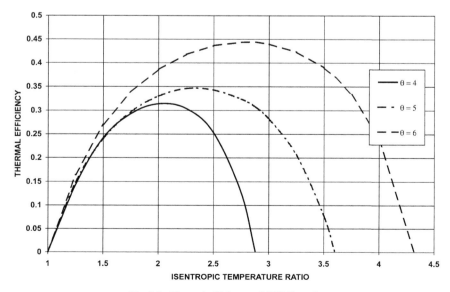

Fig. 3.9. Thermal efficiency of $[CHT]_I$ cycle.

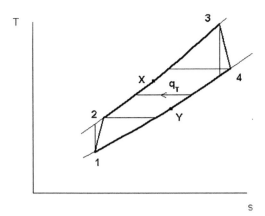

Fig. 3.10. T, s diagram for irreversible closed recuperative cycle [CHT]$_I$.

and the thermal efficiency is

$$\eta = \{\alpha[1 - (1/x)] - (x - 1)\}/\{\varepsilon\alpha[1 - (1/x)] + (1 - \varepsilon)(\beta - x)\}, \tag{3.23}$$

or

$$\eta = \{\theta[(1 - (1/x_T) - (x_C - 1)\}/\{\varepsilon\theta[1 - (1/x_T)] + (1 - \varepsilon)(\theta - x_C)\}. \tag{3.23p}$$

Optimum conditions and graphical plot

The isentropic temperature ratio for maximum efficiency (x_e) is again obtained by writing $\partial\eta/\partial x = 0$; after some algebra this yields

$$A'(x_e)^2 + B'x_e + C' = 0, \tag{3.24}$$

where

$$A' = (1 - \varepsilon)(\alpha - \beta + 1), \qquad B' = -2\alpha(1 - \varepsilon), \qquad C' = \alpha[\beta - \varepsilon(\beta + 1)].$$

For the [CHTX]$_I$ plant, with the cycle parameters quoted above for the [CHT]$_I$ plant:
(i)   with $\varepsilon = 0.5$, the values of x for maximum efficiency and maximum work become identical, $x_e = x_w = \alpha^{1/2} = 1.697$ and $\eta = 0.337$;
(ii)   with $\varepsilon = 0.75$, $x_e = 1.506$, $x_w = 1.697$, and $\eta = 0.385$.
For their graphical interpretation, Hawthorne and Davis wrote

$$\text{NDHT} = q/[c_p(T_3 - T_1)] = \varepsilon(\text{NDNW}) + [\lambda], \tag{3.25}$$

where $[\lambda] = [(2\varepsilon - 1)\text{NDCW} + (1 - \varepsilon)]$, and the efficiency as

$$\eta = \text{NDNW}/\text{NDHT} = \{\varepsilon + ([\lambda]/\text{NDNW})\}^{-1}. \tag{3.26}$$

The graphical representation is not as simple as that for the [CHT]$_I$ cycle, but still informative. It is also shown in Fig. 3.8, which gives a plot of the [CHTX]$_I$ efficiency against x for the parameters specified earlier, and for $\varepsilon = 0.75$. The term in the square brackets $[\lambda]$ in Eq. (3.25) is linear with x, passing through the x, y points $[\zeta, 0]$; $[1, (1 - \varepsilon)]$; $[(\beta + 1)/2, 1/2]$, where $\zeta = 1 - [(\beta - 1)(1 - \varepsilon)/(2\varepsilon - 1)] = -0.2$.

The effect of varying ε can also be interpreted from this type of diagram. For $\varepsilon = 1.0$, i.e. for a cycle $[CHT]_I X_R$, the maximum efficiency occurs when $r = 1.0$ (the 'square bracket' line becomes tangent to the NDNW curve at $x = 1.0$). For high values of ε (greater than 0.5), the tangent meets the curve to the left of the maximum in NDNW, whereas for low ε the tangent point is to the right. For $\varepsilon = 0.5$ the point $[\zeta, 0]$ is located at $[-\infty]$ and the 'square bracket' line becomes horizontal, touching the NDNW curve at its maximum at $r = r_w$; so that for $\varepsilon = 0.5$, $r_e = r_w$.

3.2.3. Discussion

The Hawthorne and Davis approach thus aids considerably our understanding of a/s plant performance. The main point brought out by their graphical construction is that the maximum efficiency for the simple $[CHT]_I$ cycle occurs at high pressure ratio (above that for maximum specific work); whereas the maximum efficiency for the recuperative cycle $[CHTX]_I$ occurs at low pressure ratio (below that for maximum specific work). This is a fundamental point in gas turbine design.

Fuller analyses of a/s cycles embracing intercooling and reheating were given in a comprehensive paper by Frost et al. [3], but the analysis is complex and is not reproduced here.

3.3. The $[CBT]_I$ open circuit plant—a general approach

In practical open circuit gas turbine plants with combustion, real gas effects are present (in particular the changes in specific heats, and their ratio, with temperature), together with combustion and duct pressure losses. We now develop some modifications of the a/s analyses and their graphical presentations for such open gas turbine plants, with and without heat exchangers, as an introduction to more complex computational approaches.

The Hawthorne and Davis analysis is first generalised for the $[CBT]_I$ open circuit plant, with fuel addition for combustion, f per unit air flow, changing the working fluid from air in the compressor to gas products in the turbine, as indicated in Fig. 3.11. Real gas effects are present in this open gas turbine plant; specific heats and their ratio are functions of f and T, and allowance is also made for pressure losses.

The flow of air through the compressor may be regarded as the compression of a gas with properties $(c_{pa})_{12}$ and $(\gamma_a)_{12}$ (the double subscript indicates that a mean is taken over the relevant temperature range). The work required to compress the unit mass of air in the compressor is then represented as

$$w_C = (c_{pa})_{12} T_1 [x - 1]/\eta_C, \tag{3.27}$$

where x is now given by $x = r^{(1/z)}$ and $z = (\gamma_a)_{12}/[(\gamma_a)_{12} - 1]$.

The pressure loss through the combustion chamber is allowed for by a pressure loss factor $\Delta p_{23} = (p_2 - p_3)/p_2$, so that $(p_3/p_2) = 1 - (\Delta p/p)_{23}$. Similarly, the pressure loss

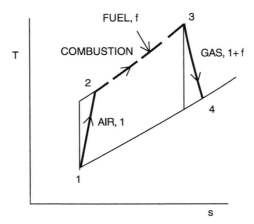

Fig. 3.11. T, s diagram for irreversible open circuit simple plant [CBT]$_I$.

factor through the turbine exhaust system is $(\Delta p/p)_{41} = (p_4 - p_1)/p_4$, and hence $(p_1/p_4) =$ $1 - (\Delta p/p)_{41}$.

The work generated by the turbine per unit mass of air after receiving combustion gas of mass $(1 + f)$ and subjected to a pressure ratio of $r[1 - (\Delta p/p)_{23})] [(1 - (\Delta p/p)_{41}]$, may then be written approximately as

$$w_T \approx (1 + f)\eta_T(c_{pa})_{12}T_3[1 - (1 + \delta)/x^n)]/n, \tag{3.28}$$

where $n = (c_{pa})_{12}/(c_{pg})_{34}$, and $\delta = \{[(\gamma)_{34} - 1]\sum(\Delta p/p)\}/(\gamma_g)_{34}$ is small.

The appearance of n as the index of x in Eq. (3.28) needs to be justified. Combustion in gas turbines usually involves substantial excess air and the molecular weight of the mixed products is little changed from that of the air supplied, since nitrogen is the main component gas for both air and products. Thus the mean gas constant (universal gas constant divided by mean molecular weight) is virtually unchanged by the combustion. It then follows that

$$(1/n) = [(\gamma_a)_{12} - 1)/(\gamma_a)_{12}][(\gamma_g)_{34} - 1)/(\gamma_g)_{34}].$$

The non-dimensional net work output (per unit mass of air) is then

$$\text{NDNW} = w/(c_{pa})_{12}(T_3 - T_1)$$

$$= \{[\alpha(1 + f)/n][1 - (1 + \delta)/x^n] - (x - 1)\}/(\beta - 1), \tag{3.29}$$

and the 'arbitrary overall efficiency' of the plant (η_O) is now defined, following Haywood [4], as

$$\eta_O = w/[-\Delta H_0], \tag{3.30}$$

where $[-\Delta H_0]$ is the change of enthalpy at temperature T_0 in isothermal combustion of a mass of fuel f with unit air flow (i.e. in a calorific value process). In the combustion process, assumed to be adiabatic,

$$[h_{a2} + fh_{f0}] = H_{g3} = (1 + f)h_{g3}, \tag{3.31}$$

where h_{f0} is the specific enthalpy of the fuel supplied at T_0.

But from the calorific value process, with heat $[-\Delta H_0] = f[CV]_0$ abstracted to restore the combustion products to the temperature T_0,

$$h_{a0} + fh_{f0} = H_{g0} + [-\Delta H_0] = (1 + f)h_{g0} + [-\Delta H_0]. \tag{3.32}$$

From Eqs. (3.31) and (3.32)

$$f[CV]_0 = (H_{g3} - H_{g0}) - (h_{a2} - h_{a0}) = (1 + f)(h_{g3} - h_{g0}) - (h_{a2} - h_{a0})$$

$$= (1 + f)(c_{pg})_{13}(T_3 - T_1) - (c_{pa})_{12}(T_2 - T_1), \tag{3.33}$$

where the ambient temperature is now taken as identical to the compressor entry temperature (i.e. $T_0 = T_1$). The non-dimensional heat supplied is, therefore

$$\text{NDHT} = f[CV]_0/[(c_{pa})_{12}(T_3 - T_1)]$$

$$= \{[(1 + f)(\beta - 1)/n'] - (x - 1)\}/(\beta - 1), \tag{3.34}$$

where $n' = (c_{pa})_{12}/(c_{pg})_{13}$.

The temperature rise in the combustion chamber may then be determined from Eq. (3.33), in the approximate form $(T_3 - T_2) = (af + b)$. Strictly a and b are functions of the temperature of the reactants and the fuel–air ratio f, but fixed values are assumed to cover a reasonable range of conditions. Accordingly, the fuel–air ratio may be expressed as

$$f = \{T_3 - T_1[1 + (x - 1)/\eta_C] - b\}/a. \tag{3.35}$$

Using this expression to determine f for given T_3 and T_1, mean values of $(\gamma_g)_{34}$ and $(c_{pg})_{34}$ for the turbine expansion may be determined from data such as those illustrated graphically in Fig. 3.12. For the weak combustion used in most gas turbines, with excess air between 200 and 400%, $f \ll 1$. Strictly, for given T_3 and T_1, the mean value of $(c_{pg})_{34}$, and indeed $(\gamma_g)_{34}$, will vary with pressure ratio.

The (arbitrary) overall efficiency may be written as

$$\eta_O = \text{NDNW/NDHT}$$

$$= \{[\alpha(1 + f)/n][1 - (1 + \delta)/x^n] - (x - 1)\}/\{[(1 + f)(\beta - 1)/n'] - (x - 1)\}. \tag{3.36}$$

Calculation of the specific work and the arbitrary overall efficiency may now be made parallel to the method used for the a/s cycle. The maximum and minimum temperatures are specified, together with compressor and turbine efficiencies. A compressor pressure ratio (r) is selected, and with the pressure loss coefficients specified, the corresponding turbine pressure ratio is obtained. With the compressor exit temperature T_2 known and T_3 specified, the temperature change in combustion is also known, and the fuel–air ratio f may then be obtained. Approximate mean values of specific heats are then obtained from Fig. 3.12. Either they may be employed directly, or n and n' may be obtained and used.

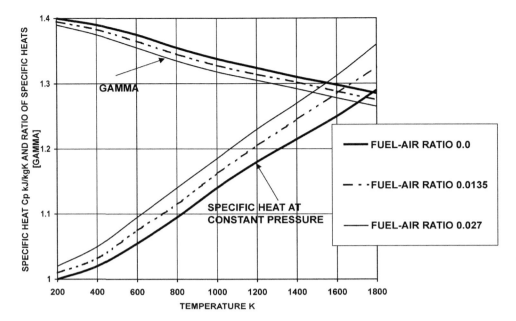

Fig. 3.12. Specific heats and their ratios for 'real' gases—air and products of combustion (after Cohen et al., see Preface [7]).

With turbine and compressor work determined, together with the 'heat supplied', the arbitrary overall efficiency is obtained.

Thus there are three modifications to the a/s efficiency analysis, involving (i) the specific heats (n and n'), (ii) the fuel–air ratio f and the increased turbine mass flow $(1 + f)$, and (iii) the pressure loss term δ. The second of these is small for most gas turbines which have large air–fuel ratios and f is of the order of 1/100. The third, which can be significant, can also be allowed for a modification of the a/s turbine efficiency, as given in Hawthorne and Davis [1]. (However, this is not very convenient as the isentropic efficiency η_T then varies with r and x, leading to substantial modifications of the Hawthorne–Davis chart.)

The first modification, involving n and n', is important and affects the Hawthorne–Davis chart. The compressor work is unchanged but the turbine work, and hence the non-dimensional net work NDNW, are increased. The heat supplied term NDHT is also changed. It should be noted here that the assumption $n' = (n + 1)/2$, used by Horlock and Woods, is not generally valid, except at very low pressure ratios.

Guha [5] pointed out some limitations in the linearised analyses developed by Horlock and Woods to determine the changes in optimum conditions with the three parameters n (and n'), f and ξ. Not only is the accurate determination of $(c_{pg})_{13}$ (and hence n') important but also the fuel–air ratio; although small, it cannot be assumed to be a constant as r is varied. Guha presented more accurate analyses of how the optimum conditions are changed with the introduction of specific heat variations with temperature and with the fuel–air ratio.

3.4. Computer calculations for open circuit gas turbines

Essentially, the analytical approach outlined above for the open circuit gas turbine plants is that used in modern computer codes. However, gas properties, taken from tables such as those of Keenan and Kaye [6], may be stored as data and then used directly in a cycle calculation. Enthalpy changes are then determined directly, rather than by mean specific heats over temperature ranges (and the estimation of n and n'), as outlined above.

A series of calculations for open circuit gas turbines, with realistic assumptions for various parameters, have been made using a code developed by Young [7], using real gas tables. These illustrate how the analysis developed in this chapter provides an understanding of, and guidance to, the performance of the real practical plants. The subscript G here indicates that the real gas effects have been included.

3.4.1. The [CBT]$_{IG}$ plant

Fig. 3.13 shows the overall efficiency for the [CBT]$_{IG}$ plant plotted against the isentropic temperature ratio for various maximum temperatures T_3 (and $\theta = T_3/T_1$, with $T_1 = 27°C$ (300 K)). The following assumptions are also made:

polytropic efficiency, $\eta_p = 0.9$ for compressor and turbine;

pressure loss fraction in combustion 0.03;

fuel (methane) and air supplied at 1 bar, 27°C (300 K).

This figure may be compared with Fig. 3.3 (which showed the a/s efficiency of plant [CHT]$_R$ as a function of x only) and Fig. 3.9 (which showed the a/s efficiency of

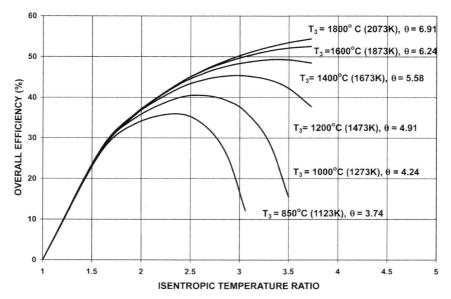

Fig. 3.13. Overall efficiency of [CBT]$_{IG}$ cycle as a function of pressure ratio r with T_3 (and temperature ratio θ) as a parameter.

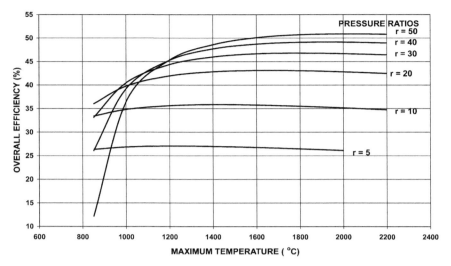

Fig. 3.14. Overall efficiency of [CBT]$_{IG}$ cycle as a function of temperature T_3 with pressure ratio r as a parameter.

plant [CHT]$_I$ as a function of x and θ). Fig. 3.13 is quite similar to Fig. 3.9, where the optimum pressure ratio increases with T_3, but the values are now more realistic.

The [CBT]$_{IG}$ efficiency is replotted in Fig. 3.14, against (T_3/T_1) with pressure ratio as a parameter. There is an indication in Fig. 3.14 that there may be a limiting maximum temperature for the highest thermal efficiency, and this was observed earlier by Horlock et al. [8] and Guha [9]. It is argued by the latter and by Wilcock et al. [10] that this is a real gas effect not apparent in the a/s calculations such as those shown in Fig. 3.9. This point will be dealt with later in Chapter 4 while discussing the turbine cooling effects.

3.4.2. Comparison of several types of gas turbine plants

A set of calculations using real gas tables illustrates the performance of the several types of gas turbine plants discussed previously, the [CBT]$_{IG}$, [CBTX]$_{IG}$, [CBTBTX]$_{IG}$, [CICBTX]$_{IG}$ and [CICBTBTX]$_{IG}$ plants. Fig. 3.15 shows the overall efficiency of the five plants, plotted against the overall pressure ratio (r) for $T_3 = 1200°C$. These calculations have been made with assumptions similar to those made for Figs. 3.13 and 3.14. In addition (where applicable), equal pressure ratios are assumed in the LP and HP turbomachinery, reheating is set to the maximum temperature and the heat exchanger effectiveness is 0.75.

The first point to note is that the classic Hawthorne and Davis argument is reinforced— that the optimum pressure ratio for the [CBT]$_{IG}$ plant ($r \approx 45$) is very much higher than that for the [CBTX]$_{IG}$ plant ($r \approx 9$). (The optimum r for the latter would decrease if the effectiveness (ε) of the heat exchanger were increased, but it would increase towards that of the [CBT]$_{IG}$ plant if ε fell towards zero.)

While the lowest and highest optimum pressure ratios are for these two plants, the addition of reheating and intercooling increases the optimum pressure ratios above that of

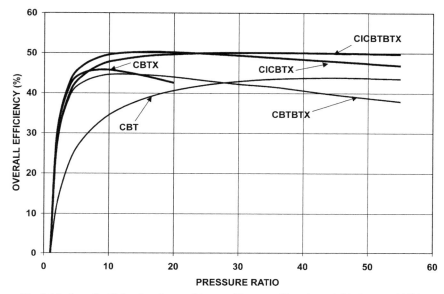

Fig. 3.15. Overall efficiencies of several irreversible gas turbine plants (with $T_{max} = 1200°C$).

the simple recuperative plant. The highest efficiency (with a high optimum pressure ratio) occurs for the most complex $[CICBTBTX]_{IG}$ plant, but the graph of efficiency (η) with pressure ratio is very flat at the high pressure ratios, of $30-55$ (η approaches the efficiency of a plant with heat supplied at maximum temperature and heat rejected at minimum temperature).

Finally, carpet plots of efficiency against specific work are shown in Fig. 3.16, for all these plants. The increase in efficiency due to the introduction of heat exchange, coupled with reheating and intercooling, is clear. Further the substantial increases in specific work associated with reheating and intercooling are also evident.

3.5. Discussion

The discussion of the performance of gas turbine plants given in this chapter has developed through four steps: reversible a/s cycle analysis; irreversible a/s cycle analysis; open circuit gas turbine plant analysis with approximations to real gas effects; and open circuit gas turbine plant computations with real gas properties. The important conclusions are as follows:

(a) The initial conclusion for the basic Joule–Brayton reversible cycle $[CHT]_R$, that thermal efficiency is a function of pressure ratio (r) only, increasing with r, is shown to have major limitations. The introduction of irreversibility in a/s cycle analysis shows that the maximum temperature has a significant effect; thermal efficiency increases with (T_3/T_1), and so does the optimum pressure ratio for maximum efficiency.

(b) The a/s analyses show quite clearly that the introduction of a heat exchanger leads to higher efficiency at low pressure ratio, and that the optimum pressure ratio for the

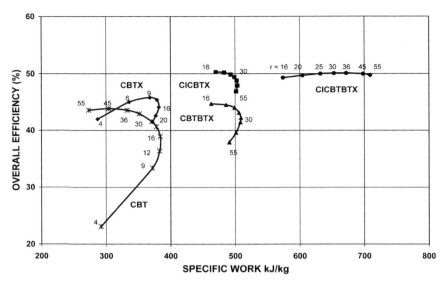

Fig. 3.16. Overall efficiency and specific work of several irreversible gas turbine plants (with $T_{max} = 1200°C$).

[CHTX]$_I$ cycle is much lower than that of the [CHT]$_I$ cycle. The optimum pressure ratio for maximum specific work falls between these two pressure ratios.

(c) The major benefits of the addition of reheating and intercooling to the unrecuperated plants are to increase the specific work. However, when these features are coupled with heat exchange the full benefits on efficiency are obtained.

References

[1] Hawthorne, W.R. and Davis, G.de V. (1956), Calculating gas turbine performance, Engineering 181, 361–367.

[2] Horlock, J.H. and Woods, W.A. (2000), Determination of the optimum performance of gas turbines, Proc. Instn. Mech. Engrs. J. Mech. Engng. Sci. 214(C), 243–255.

[3] Frost, T.H., Agnew, B. and Anderson, A. (1992), Optimisation for Brayton–Joule gas turbine cycles, Proc. Instn. Mech. Engrs. Part A, J. Power Energy 206(A4), 283–288.

[4] Haywood, R.W. (1991), Analysis of Engineering Cycles, 4th edn, Pergamon Press, Oxford.

[5] Guha, A. (2003), Effect of internal combustion and real gas properties on the optimum performance of gas present status turbines, Instn. Mech. Engrs., in press.

[6] Keenan, J.H. and Kaye, J. (1945), Gas Tables, Wiley, New York.

[7] Young, J.B. (1998), Computer-based Project on Combined-Cycle Power Generation. Cambridge University Engineering Department Report.

[8] Horlock, J.H., Watson, D.T. and Jones, T.V. (2001), Limitations on gas turbine performance imposed by large turbine cooling flows, ASME J. Engng Gas Turbines Power 123(3), 487–494.

[9] Guha, A. (2000), Performance and optimization of gas turbines with real gas effects, Proc. Instn. Mech. Engnrs. Part A 215, 507–512.

[10] Wilcock, R.C., Young, J.B. and Horlock, J.H. (2002), Real Gas Effects on Gas Turbine Plant Efficiency. ASME Paper GT-2002-30517.

Chapter 4

CYCLE EFFICIENCY WITH TURBINE COOLING (COOLING FLOW RATES SPECIFIED)

4.1. Introduction

It was pointed out in Chapter 1 that the desire for higher maximum temperature (T_{max}) in thermodynamic cycles, coupled with low heat rejection temperature (T_{min}), is essentially based on attempting to emulate the Carnot cycle, in which the efficiency increases with (T_{max}/T_{min}).

It has been emphasised in the earlier chapters that the thermal efficiency of the gas turbine increases with its maximum nominal temperature, which was denoted as T_3. Within limits this statement is true for all gas turbine-based cycles and can be sustained, although not indefinitely, as long as the optimum pressure ratio is selected for any value of T_3; further the specific power increases with T_3. However, in practice higher maximum temperature requires improved combustion technology, particularly if an increase in harmful emissions such as NO_x is to be avoided.

Thus, the maximum temperature is an important parameter of overall cycle performance. But for modern gas turbine-based systems, which are cooled, a precise definition of maximum temperature is somewhat difficult, and Mukherjee [1] suggested three possible definitions. The first is the combustor outlet temperature (T_{cot}) which is based on the average temperature at exit from the combustion chamber. However, in a practical system, this does not take into account the effect of cooling flows that are introduced subsequently (e.g. in the first turbine row of guide vanes). So a second definition involving the rotor inlet temperature (T_{rit}) has tended to be used more widely within the gas turbine industry. T_{rit} is based on the averaged temperature taken at the exit of the first nozzle guide vane row, NGV (i.e. at entry to the first rotor section), and this can be calculated assuming that the NGV cooling air completely mixes with the mainstream. A third definition, the so-called ISO firing temperature, T_{ISO}, can be calculated from the combustion equations and a known fuel–air ratio, but this definition is less frequently used (it should theoretically yield the same temperature as T_{cot}).

T_{cot} and T_{rit} are both important in the understanding of relative merits of candidate cooling systems, and we shall later emphasise the difference between T_{cot} and T_{rit}. Without improvements in materials and/or heat transfer, it is doubtful whether much higher T_{rit} values can be achieved in practice; as a result, a practical limit on plant efficiency may be near, before the stoichiometric limit is reached. Below we refer to T_{cot} as T_3, the maximum temperature in cycle analyses, and T_{rit} as T_5, the temperature after cooling of the first NGV row.

In this chapter, cycle calculations are made with assumed but realistic estimates of the probable turbine cooling air requirements which include some changes from the uncooled thermal efficiencies. Indeed it is suggested that for modern gas turbines there may be a limit on the combustion temperature for maximum thermal efficiency [2,3].

As discussed in Chapter 3, analysis of *uncooled* gas turbine cycles was developed in three stages:

(a)   for air-standard (a/s) reversible cycles;
(b)   for a/s irreversible cycles;
(c)   for real gas irreversible cycles.

By introducing the effects of turbine cooling a similar development is followed in this chapter. Here, we look initially at the effect of turbine cooling in

(a)   in reversible a/s cycles; and
(b)   in irreversible a/s cycles.

For the purpose of the cycle analyses (a) and (b), the following assumptions are made: (i) cooling is of the open type, with a known air flow fraction (ψ) first cooling a blade row and then mixing with the mainstream; and (ii) complete mixing takes place, under adiabatic conditions, at constant static pressure and low Mach number (and therefore constant stagnation pressure). Before moving on to more realistic cycle calculations (but with the cooling air quantity (ψ) assumed to be known), we consider the irreversibilities in the turbine cooling process, showing how changes in stagnation pressure and temperature (and entropy) are related to ψ. These changes are then used in cycle calculations for which ψ is again specified, but real gas effects and stagnation pressure losses are included.

Subsequently, in Chapter 5, we shall show how the cooling quantities may be determined; we give even more practical cycle calculations, with these cooling quantities (ψ) being determined practically rather than specified ab initio. But for the discussions in this chapter, in which we assess how important cooling is in modifying the overall thermodynamics of gas turbine cycle analysis, it is assumed that ψ is known.

The nomenclature introduced by Hawthorne and Davis [4] is adopted and; gas turbine cycles are referred to as follows: CHT, CBT, CHTX, CBTX, where C denotes compressor; H, air heater; B, burner (combustion); T, turbine; X, heat exchanger. R and I indicate reversible and irreversible. The subscripts U and C refer to uncooled and cooled turbines in a cycle, and subscripts 1, 2, M indicate the number of cooling steps (one, two or multi-step cooling). Thus, for example, $[CHT]_{IC2}$ indicates an irreversible cooled simple cycle with two steps of turbine cooling. The subscript T is also used to indicate that the cooling air has been throttled from the compressor delivery pressure.

4.2. Air-standard cooled cycles

The initial analysis [5] is presented by reference to closed a/s cycles using a perfect gas as a working fluid in an externally heated plant. As for the uncooled cycles studied in Chapter 3, it is argued subsequently that many of the conclusions reached in this way remain substantially valid for open cycles with combustion, i.e. for those involving real gases with variable composition and specific heats varying with temperature.

The arguments of this section are developed sequentially, starting with internally reversible cycles and then considering irreversibilities. Here we concentrate on the gas turbine with simple closed or open cycle (CHT, CBT).

4.2.1. Cooling of internally reversible cycles

4.2.1.1. Cycle [CHT]$_{RC1}$ with single step cooling

Consider first a cycle with reversible compression and expansion, but one in which, after a unit flow of the compressed gas has been heated externally, it is cooled by mixing with the remaining compressor delivery air (ψ) before entering the turbine in the internal cycle, which is otherwise reversible (Fig. 4.1 shows the T, s chart). This single step of cooling is representative of cooling the nozzle guide vanes of the first stage in a real gas turbine plant, reducing the rotor inlet temperature from $T_3 = T_{\text{cot}}$ to $T_5 = T_{\text{rit}}$.

We assume low velocity (constant pressure) mixing of the 'extra' cooling gas mass flow (ψ) at absolute temperature T_2 with the gas stream (of unit mass flow), which has been heated to the maximum temperature $T_3 = T_B$. From the steady flow energy equation, if both streams have the same specific heat (c_p), it follows that

$$\psi T_2 + T_3 = (1 + \psi)T_5, \tag{4.1}$$

where T_5 is the resulting temperature in the mixed stream, before it is expanded through the turbine. The turbine work output is now $W_T = (1 + \psi)c_p T_5[(1 - (1/x)]$, and the

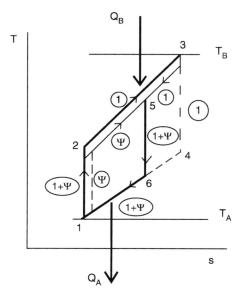

Fig. 4.1. Temperature–entropy diagram for single-step cooling—reversible cycle [CHT]$_R$ (after Ref. [5]).

compressor work is $W_C = (1 + \psi)c_p T_1(x - 1)$. But the heat supplied, before the mixing process, to the stream of unit mass flow is still $Q_B = c_p(T_3 - T_2)$, which from Eq. (4.1) may be written as

$$Q_B = (1 + \psi)c_p(T_5 - T_2). \tag{4.2}$$

Hence, the internal thermal efficiency is

$$
\begin{aligned}
(\eta)_{RC1} &= (W_T - W_C)/Q_B \\
&= \{(1 + \psi)c_p T_5[1 - (1/x)] - (1 + \psi)c_p T_1(x - 1)\}/\{(1 + \psi)c_p(T_5 - T_2)\} \\
&= [(\theta'/x) - 1](x - 1)/[(\theta' - 1) - (x - 1)], \tag{4.3}
\end{aligned}
$$

where $\theta' = T_5/T_1$. But this expression can be simplified as

$$(\eta)_{RC1} = [1 - (1/x)] = (\eta)_{RU}, \tag{4.4}$$

which is independent of θ'.

Thus the cooled 'reversible' cycle [CHT]$_{RC1}$ with a first rotor inlet temperature, T_5, will have an internal thermal efficiency exactly the same as that of the uncooled cycle [CHT]$_{RU}$ with a higher turbine entry temperature $T_3 = T_B$, and the same pressure ratio. There is no penalty on efficiency in cooling the turbine gases at entry; but note that the specific work output, $w = (w_T - w_C)/c_p T_1 = [(\theta'/x) - 1](x - 1)$, is reduced, since $\theta' < \theta$.

This result requires some explanation. An argument was given by Denton [6], who pointed out that the expansion of the mixed gas $(1 + \psi)$ from T_5 to T_6 may be considered as a combination of unit flow through the turbine from T_3 to T_4, and an expansion of a flow of ψ from T_2 to T_1, through a 'reversed' compressor (Fig. 4.2). The cycle [1,2,3,5,6,1] of Fig. 4.2a is equivalent to two parallel cycles as indicated in Fig. 4.2b: a cycle [1,2,3,4,1] with unit circulation; plus another cycle passing through the state points [1,2,2,1] with a circulation ψ. The second cycle has the same efficiency as the first (but vanishingly small work output) so that the combined cooled cycle has the same efficiency as each of the two

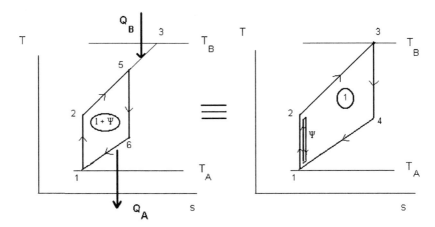

Fig. 4.2. Temperature–entropy diagram for single-step cooling—equivalent two cycles (after Ref. [5]).

component cycles. This interpretation will also be useful when we consider the internally irreversible cycles later.

There is an apparent paradox here that as the cooled cycle contains an irreversible process (constant pressure mixing), its efficiency might be expected to be lower than the original uncooled cycle. The answer to this paradox follows from consideration of all the irreversibilities in the cycle and we refer back to the analysis of Section 3.2.1.1, for the rational efficiency of the $[CHT]_{RU}$ cycle. The irreversibility associated with the heat supply is unchanged, as given in Eq. (3.3), but the irreversibility associated with the heat rejection Q_A between temperatures T_6 and $T_1 = T_A$ becomes

$$I_A = Q_A - T_A \int (dQ_A/T) = Q_A - (1 + \psi)c_p T_A \ln(T_6/T_A). \tag{4.5}$$

The irreversibility in the adiabatic mixing is

$$I_M = T_A[(1 + \psi)s_5 - s_3 - \psi s_2] = c_p T_A[[\psi \ln(T_5/T_2)] - \ln(T_B/T_5)], \tag{4.6}$$

since low Mach number and constant pressure mixing have been assumed.

The sum of the irreversibilities I_A and I_M is thus

$$I_A + I_M = Q_A - T_A c_p \ln[(T_6/T_A)(T_B/T_5)] + \psi c_p T_A \ln[(T_5/T_2)(T_A/T_6)]. \tag{4.7}$$

But, since $T_B/T_4 = T_5/T_6 = T_2/T_A = x$, this equation becomes

$$I_B + I_M = Q_A - c_p T_A \ln(T_4/T_A), \tag{4.8}$$

which is the same as the irreversibility associated with heat rejection in the uncooled cycle $[CHT]_{RU}$ given in Chapter 3, Eq. (3.4). Further the maximum work, W_{max}, is unchanged from that given in the $[CHT]_{RU}$ cycle, as is the rational efficiency. The sum of all the irreversibilities are the same in the two cycles, $[CHT]_{RU}$ and $[CHT]_{RC}$, but they are broken down and distributed differently. This point is amplified by Young and Wilcock [7].

4.2.1.2. Cycle $[CHT]_{RC2}$ with two step cooling

A reversible cycle with turbine expansion split into two steps (high pressure, HP, and low pressure, LP) is illustrated in the T, s diagram of Fig. 4.3. The mass flow through the heater is still unity and the temperature rises from T_2 to $T_3 = T_B$; hence the heat supplied Q_B is unchanged, as is the overall isentropic temperature ratio (x). But cooling air of mass flow ψ_H is used at entry to the first HP turbine (of isentropic temperature ratio x_H) and additional cooling of mass flow ψ_L is introduced subsequently into the LP turbine (of isentropic temperature ratio x_L). The total cooling flow is then $\psi = \psi_H + \psi_L$.

As is shown in Fig. 4.3a, the lower pressure cooling is fed by air ψ_L at state 7, at a corresponding pressure p_7 and a temperature T_7, and this mixes with air $(1 + \psi_H)$ from the HP exhaust at temperature T_9 to produce a temperature T_8 as indicated in the diagram. The full turbine gas flow $(1 + \psi)$ then expands through a pressure ratio x_L to a temperature T_{10}, and subsequently rejects heat, finishing at $T_1 = T_A$.

But this expansion through the LP turbine may be considered as two parallel expansions. The first is of mass flow $(1 + \psi_H)$ from the temperature T_9 to a temperature T_6 (a continuation of the expansion of $(1 + \psi_H)$ from 5 to 9); and the second is of mass flow ψ_L through a reversed compressor from state 7 to state 1 (which cancels out the

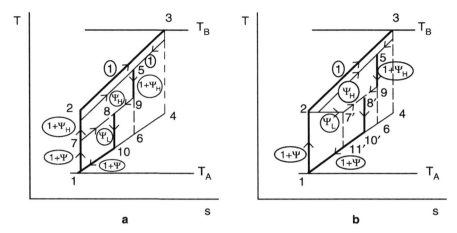

Fig. 4.3. Temperature–entropy diagram for two step cooling. (a) Cooling air taken at appropriate pressures and (b) LP cooling air throttled from compressor exit.

compression of ψ_L from 1 to 7). An equivalent cycle of mass flow $(1 + \psi_H)$ through the states [1,2,5,6] is thus produced, with the state 5 formed after mixing of (unit) heated gas at temperature T_3 with cooling air ψ_H at temperature T_2. But the efficiency of that cycle [1,2,5,6] is the same as that of the original uncooled cycle [1,2,3,4], with a unit mass flow. Thus, the original conclusion that single step cooling does not change the efficiency of a reversible simple cycle [CHT]$_{RU}$, is extended; two step cooling, with air abstracted from the compressor at the appropriate pressure, also does not change the thermal efficiency,

$$(\eta)_{RC2} = [1 - (1/x)] = (\eta)_{RU}. \tag{4.9}$$

However, it is important to note that this conclusion becomes invalid if the air for cooling the LP turbine is taken from compressor delivery (as in Fig. 4.3b) and then throttled at constant temperature ($T_2 = T_{7'}$) to the lower pressure before being mixed with the gas leaving the HP turbine. The thermal efficiency drops as another internal irreversibility is introduced; it can be shown [5] that

$$(\eta)_{RC2T} = (\eta)_{RU} - [\psi_L(x_H - 1)]/(\theta - x). \tag{4.10}$$

The drop in thermal efficiency due to throttling the LP air is very small. For example, a cycle [CHT]$_{RC2}$ with a pressure ratio of $r = 36.27$ ($x = 2.79$) has a thermal efficiency of $(\eta)_{RC2} = (\eta)_{RU} = 0.642$. For the cycle [CHT]$_{RC2T}$ with $\psi_L = 0.05$ and $x_H = 1.22$, $\theta = 6$, the second term in Eq. (4.10) is only 0.003, i.e. the thermal efficiency drops from 0.642 to $(\eta)_{RC2T} = 0.639$.

4.2.1.3. Cycle [CHT]$_{RCM}$ with multi-step cooling

The argument developed in Section 4.2.1.2 can be extended for three or more steps of cooling, to give the same efficiency as the uncooled cycle. Indeed the efficiency will be the same for multi-step cooling, with infinitesimal amounts of air abstracted at an infinite number of points along the compressor to cool each infinitesimal turbine stage at the required pressures.

But another approach to multi-step cooling [8, 9] involves dealing with the turbine expansion in a manner similar to that of analysing a polytropic expansion. Fig. 4.4 shows gas flow $(1 + \psi)$ at (p, T) entering an elementary process made up of a mixing process at constant pressure p, in which the specific temperature drops from temperature T to temperature T', followed by an isentropic expansion in which the pressure changes to $(p + \mathrm{d}p)$ and the temperature changes from T' to $(T + \mathrm{d}T)$.

In the first mixing process, the entry mainstream flow $(1 + \psi)$ mixes with cooling flow $\mathrm{d}\psi$ drawn from the compressor at temperature T_{comp}. Thus, if c_p is constant, then

$$(1 + \psi + \mathrm{d}\psi)c_p T' = (1 + \psi)c_p T + \mathrm{d}\psi c_p T_{\mathrm{comp}},$$

and

$$c_p(T - T') = c_p(T' - T_{\mathrm{comp}})\mathrm{d}\psi/(1 + \psi). \tag{4.11}$$

In the second process of isentropic expansion

$$c_p[(T + \mathrm{d}T) - T'] = v\mathrm{d}p, \tag{4.12}$$

where v is the specific volume.

Subtracting Eq. (4.11) from Eq. (4.12), it then follows that in the overall elementary process, $(p, T, 1 + \psi)$ to $(p + \mathrm{d}p, T + \mathrm{d}T, 1 + \psi + \mathrm{d}\psi)$,

$$c_p\mathrm{d}T + c_p(T' - T_{\mathrm{comp}})\mathrm{d}\psi/(1 + \psi) = v\mathrm{d}p, \tag{4.13}$$

or

$$c_p\mathrm{d}T/T = R\mathrm{d}p/p - c_p(T' - T_{\mathrm{comp}})\mathrm{d}\psi/[T(1 + \psi)]. \tag{4.14}$$

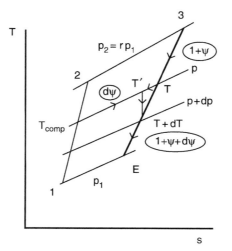

Fig. 4.4. Temperature–entropy diagram for multi-step cooling—reversible cycle [CHT]$_{\mathrm{RCM}}$ (after Ref. [5]).

There are two approaches to integrating this equation:
(a) the three terms can be integrated separately to give a p, T, ψ relation; and
(b) two of the three terms can be brought together if an expression for $d\psi/dT$ is known; a more familiar polytropic p, T type of relation can then be obtained.

In the first approach $[T' - T_{comp}]/T$ in Eq. (4.14) may be written approximately as $[1 - (T_{comp}/T)]$, for a process which does not deviate too far from the original (uncooled) isentropic expansion; and further (T_{comp}/T) may be approximated to $T_2/T_3 = x/\theta$. Then Eq. (4.14) may be integrated to give

$$T/p^{(\gamma-1)/\gamma} = C/[1 + \psi]^\delta, \tag{4.15}$$

where $\delta = 1 - (x/\theta)$. The cooling is carried out over the full turbine expansion, to an exit state (p_1, T_E), so

$$\theta_E = T_E/T_1 = (\theta/x)/[1 + \psi_E]^\delta < (\theta/x). \tag{4.16}$$

In the second approach, a value for ψ_E is not assumed but a relationship for $d\psi/dT$ is determined from semi-empirical expressions for the amount of cooling air that is required in an (elementary) turbine blade row. One such relationship, derived in Ref. [5], gives

$$\delta d\psi/[1 + \psi] = -\lambda dT/T, \tag{4.17}$$

where $\lambda = 2Cw^+[1 - (x/\theta)]/[\Phi(\gamma - 1)M_u^2] = 2Cw^+\delta/[\Phi(\gamma - 1)M_u^2]$, in which C and w^+ are parameters obtained from the definition of the blade cooling effectiveness, M_u is the blade Mach number and $\Phi = c_p\Delta T/U^2$ is the stage loading coefficient, with ΔT the (positive) temperature drop across the stage.

Eq. (4.14) can then be integrated to give

$$T/p^\sigma = \text{constant}, \tag{4.18}$$

where $\sigma = (\gamma - 1)/\gamma(1 - \lambda)$ and it follows that

$$\theta_E = T_E/T_1 = \theta/r^\sigma. \tag{4.19}$$

4.2.1.4. The turbine exit condition (for reversible cooled cycles)

There is a link between the thermal efficiency and the turbine exit temperature T_E. It results from expressing the thermal efficiency of the cycle in the form

$$\eta = [1 - Q_A/Q_B] = 1 - (1/x), \tag{4.20}$$

and it has been argued that this equation is valid for all the reversible cycles considered above (except for the second step cooling by throttled compressor delivery air, $[CHT]_{RC2T}$).

The heat supplied is $Q_B = c_p[T_3 - T_2]$, and for each of these reversible cycles the heat rejected will be $Q_A = c_p(1 + \psi_E)(T_E - T_1)$, Thus, the efficiency is given by

$$\eta = [1 - (Q_A/Q_B)] = 1 - [(1 + \psi_E)(T_E - T_1)/(T_3 - T_2)] = 1 - (1/x), \tag{4.21}$$

where ψ_E is the total amount of cooling air supplied from the compressor. The exhaust temperature T_E is therefore a function of ψ_E and, if ψ_E^2 is neglected, then it is given by

$$\theta_E = T_E/T_1 = 1 + [(\theta - x)/x(1 + \psi_E)] \approx (\theta/x)(1 - \psi_E) + \psi_E. \qquad (4.22)$$

This expression for θ_E can also be obtained directly from Eqs. (4.16) and (4.19) [5].

4.2.2. Cooling of irreversible cycles

From the study of uncooled cycles in Chapter 3, we next move to consider irreversible cycles with compressor and turbine isentropic efficiencies, η_C and η_T, respectively.

The a/s efficiency of the irreversible uncooled cycle [CHT]$_{IU}$ was given in Eq. (3.13) as

$$(\eta)_{IU} = [(\alpha - x)(x - 1)]/[x(\beta - x)], \qquad (4.23)$$

where $\alpha = \eta_C \eta_T \theta$ and $\beta = 1 + \eta_C(\theta - 1)$, with $\theta = T_3/T_1$, and this will be used as a comparator for the modified (cooled) cycles. As a numerical illustration, with $T_3 = 1800$ K, $T_1 = 300$ K ($\theta = 6.0$), $\eta_T = 0.9$, $\eta_C = 0.8$, $\alpha = 4.32$, and $\beta = 5$, the uncooled thermal efficiency $(\eta)_{IU}$ is a maximum of 0.4442, at $x = 2.79$ ($r = 36.27$), compared with the reversible efficiency, $(\eta)_{RU} = 0.642$. The expression for efficiency, Eq. (4.23), is modified when turbine cooling takes place.

4.2.2.1. Cycle with single-step cooling [CHT]$_{IC1}$
Consider again the simplest case of compressor delivery air (mass flow ψ, at T_2), mixed at constant pressure with unit mass flow of combustion products (at T_3) to give mass flow $(1 + \psi)$ at T_5 (see the T, s diagram of Fig. 4.5). The compression and expansion processes are now irreversible.

Again, following Denton [6], the turbine expansion from T_5 to T_6 may be interpreted as being equivalent to an expansion of unit flow from T_3 to T_4 together with an expansion

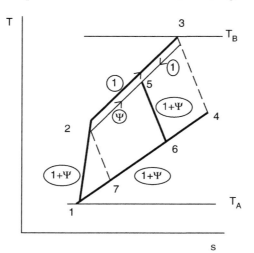

Fig. 4.5. Temperature–entropy diagram for single-step cooling—irreversible cycle [CHT]$_{IC1}$ (after Ref. [5]).

of gas flow ψ from T_2 to T_7. However, the work input to compress the fraction ψ of the mainstream compressor flow is not now effectively cancelled by the latter expansion. The cycle [1,2,3,5,6,1] is thus equivalent to a combination of two cycles: one of unit mass flow following the original uncooled cycle state points [1,2,3,4,1] (and with the same efficiency $(\eta)_{IU}$); and another of mass flow ψ following the state points [1,2,2,7,1]. The second cycle effectively has a negative work output and a heat supply which in the limit is zero.

Analytically, the efficiency of this combination of the two cycles may be expressed as

$$(\eta)_{IC1} = \{(T_3 - T_4) + \psi(T_2 - T_7) - (1 + \psi)(T_2 - T_1)\}/(T_3 - T_2)$$

$$= (\eta)_{IU} - \psi(T_7 - T_1)/(T_3 - T_2) = (\eta)_{IU} - \psi\varepsilon(x - 1)/(\beta - x), \qquad (4.24)$$

where $\varepsilon = [1 - (\eta_T\eta_C/x) - \eta_T + (\eta_T/x)]$.

Thus the efficiency of the cooled cycle is now less than that of the uncooled cycle by an amount which is directly proportional to the cooling air used (ψ). The magnitude of the correction term to the uncooled efficiency is small for a cycle with compressors and turbines of high isentropic efficiency. For a cooled version of the uncooled cycle considered earlier, with $\psi = 0.15$ and $x = 2.79$, the second term on the right hand side of Eq. (4.24) is 0.0200, the efficiency dropping from $(\eta)_{IU} = 0.4442$ to $(\eta)_{IC1} = 0.4242$. Thus cooling apparently has a relatively small effect on cycle efficiency, even when the amount of cooling flow needed becomes quite large. But Eq. (4.24) indicates that for a given ψ the reduction in efficiency should also decrease as the maximum temperature increases, for a given pressure ratio.

4.2.2.2. Efficiency as a function of combustion temperature or rotor inlet temperature (for single-step cooling)

An important point needs to be re-emphasised, that the cooled efficiency $(\eta)_{IC1}$ with 'combustion' temperature ($T_3 = T_{cot}$) is the same as the uncooled efficiency $(\eta)_{IU}$ at the 'rotor inlet temperature' ($T_5 = T_{rit}$)

Fig. 4.6 shows diagrammatically both $(\eta)_{IU}$ and $(\eta)_{IC1}$ plotted against maximum temperature (in Fig. 4.6a). The efficiency of the cooled gas turbine $(\eta)_{IC1}$ (point A) is less

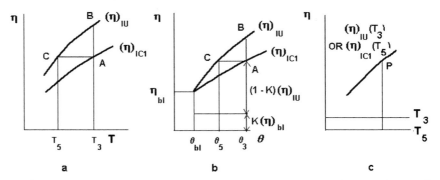

Fig. 4.6. Efficiency plots for irreversible uncooled and single-step cooled cycles (after Ref. [5]). (a) Efficiency against maximum temperature. (b) Efficiency against non-dimensional maximum temperature. (c) Efficiency against combustion temperature (T_3) and rotor inlet temperature (T_5).

than the efficiency of the uncooled turbine $(\eta)_{IU}$ at the same T_3 (point B), as given in Eq. (4.24). But it is the same as the efficiency of the uncooled turbine $(\eta)_{IU}$ at point C, at a maximum temperature T_5 (the rotor inlet temperature of the cooled turbine). Here the analysis of Section 4.2.2.1, for a/s cycles with constant specific heats, is developed further, to find the slopes of the curves $(\partial \eta / \partial \theta)_x$ at all the three points A, B and C; the slopes are then used to determine the relations between the expressions for $(\eta)_{IC1}$ and $(\eta)_{IU}$.

An approximate relation for the cooling fraction ψ obtained by El-Masri [10], and derived in Appendix A, is also used,

$$\psi = K[T_3 - T_{bl}]/[T_{bl} - T_2], \tag{4.25}$$

where T_{bl} is the allowable (constant) blade temperature, T_2 is the compressor delivery (coolant) temperature and K is a constant (approximately 0.05). Differentiation of Eq. (4.24) at constant x (and T_2), and using Eq. (4.25), yields

$$[\partial(\eta)_{IC1}/\partial \theta]_x = \varepsilon \eta_C (x - 1)/(\beta - x)^2 + \varepsilon \psi(x - 1)\eta_C \{1 - [(\beta - x)/\tau \eta_C]\}/(\beta - x)^2, \tag{4.26}$$

where $\tau = (\theta_3 - \theta_{bl})$, with $\theta_{bl} = T_{bl}/T_1$ assumed constant. The first term on the right hand side gives the rate of increase of thermal efficiency in the absence of cooling, $[\partial(\eta)_{IU}/\partial \theta]_x$. After some algebra [5], it follows that

$$\{[\partial(\eta)_{IC1}/\partial \theta]_x\}_A = (1 - K)\{[\partial(\eta)_{IU}/\partial \theta]_x\}_B. \tag{4.27}$$

So in Fig. 4.6a, the slope of the $(\eta)_{IC1}$ curve at A is $(1 - K)$ times the slope of the $(\eta)_{IU}$ curve at B. $(\eta)_{IC1}$ thus increases with T_3 at a smaller rate than $(\eta)_{IU}$. Eq. (4.27) may then be integrated to (non-dimensional) temperature θ, from θ_{bl} where $\psi_{bl} = 0$, and the uncooled and cooled efficiencies are the same, $[(\eta)_{IU}]_{bl} = [(\eta)_{IC1}]_{bl} = (\eta)_{bl}$.

Thus

$$(\eta)_{IC1} - (\eta)_{bl} = (1 - K)[(\eta)_{IU} - (\eta)_{bl}], \tag{4.28}$$

or

$$(\eta)_{IC1} = (1 - K)(\eta)_{IU} + K(\eta)_{bl}, \tag{4.29}$$

as illustrated in Fig. 4.6b. Hence

$$\Delta\eta = (\eta)_{IU} - (\eta)_{IC1} = K[(\eta)_{IU} - (\eta)_{bl}]. \tag{4.30}$$

An alternative approach is shown in Fig. 4.6c. The cooled efficiency $(\eta)_{IC1}$ may be presented as a unique function of the rotor inlet temperature (T_5), for a given x and component efficiencies. But from the El-Masri expression, Eq. (4.25), it can be deduced that the cooling air quantity ψ is a function of the combustion temperature T_3, for a given x (and T_2) and a selected blade temperature T_{bl}, so that from the steady flow energy equation for the mixing process, there is a value of T_3 corresponding to the rotor inlet temperature T_5. Analytically, we may therefore state that $(\eta)_{IC1} = f(T_5)$ and $(\eta)_{UC} = f(T_3)$, but taking note that $T_5 = f(T_3)$. Thus, uncooled and cooled efficiencies may be plotted against two horizontal scales, T_3 and T_5, as indicated in Fig. 4.6c, which results in a single line. This point is further discussed in Section 4.4.

4.2.2.3. Cycle with two step cooling [CHT]$_{IC2}$

For two step cooling, now with irreversible compression and expansion, Fig. 4.7 shows that the turbine entry temperature is reduced from T_3 to T_5 by mixing with the cooling air ψ_H taken from the compressor exit, at state 2, pressure p_2, temperature T_2 (Fig. 4.7a). After expansion to temperature T_9, the turbine gas flow $(1 + \psi_H)$ is mixed with compressor air at state 7 (mass flow ψ_L) abstracted at the same pressure p_7 with temperature T_7, to give a cooled gas flow $(1 + \psi_H + \psi_L)$ at temperature T_8. This gas is then expanded to temperature T_{10}.

It may be shown [5] that

$$\eta_{IC2} = (\eta)_{IU} - [\psi_H \varepsilon(x - 1) + \psi_L \varepsilon_L (x_L - 1)]/(\beta - x), \tag{4.31}$$

where $\varepsilon_L = [1 - (\eta_T \eta_C/x_L) - \eta_T + (\eta_T/x_L)]$. It has been assumed here that $x \gg x_H$, so that the efficiencies η_C and η_T are the same over the isentropic temperature ratios x and x_L.

For the a/s example quoted earlier, with this form of two stage cooling (with $x = 2.79$, $x_H = 1.22$, $\psi_H = 0.1$, $\psi_L = 0.05$), the thermal efficiency is reduced from 0.4442 (uncooled) to 0.4257, i.e. by 0.0185, still not a significant reduction. If the second step of cooling uses compressor delivery air rather than air taken at the appropriate pressure along the compressor, then the analysis proceeds as before, except that the expansion work for the processes 7, 11 in Fig. 4.7a is replaced by that corresponding to $7'$, $11'$ in Fig. 4.7b. It may be shown [5] that the efficiency may then be written as

$$\eta_{IC2} = (\eta)_{IU} - \{\psi \varepsilon(x - 1) - \psi_L(\varepsilon - \varepsilon_L) - \eta_T \psi_L(x_H - 1)\}/(\beta - x). \tag{4.32}$$

The second term in the curly brackets is very small indeed and may be ignored; the last term in these brackets effectively represents the throttling loss in this irreversible cycle.

For the numerical example the cooled efficiency becomes 0.4205, a reduction of 0.0237 from $(\eta)_{RU} = 0.4442$. The extra loss in efficiency for throttling the cooling air from compressor discharge to the appropriate pressure at the LP turbine entry is thus 0.0052 for the numerical example, which is again quite small.

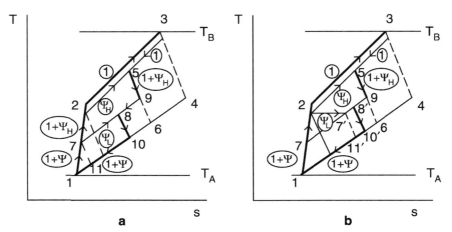

Fig. 4.7. Temperature–entropy diagram for two step cooling—irreversible cycle. (a) Cooling air taken at appropriate pressures. (b) Cooling air throttled from compressor exit (after Ref. [5]).

4.2.2.4. *Cycle with multi-step cooling* [CHT]$_{ICM}$

The two step cooling example given above can in theory be extended to multi-step cooling of the turbine. It is more convenient to treat the turbine expansion as a modification of normal polytropic expansion; the analysis is essentially an adaptation of that given in Section 4.2.1.3 for the multi-step cooled turbine cycle.

If the polytropic efficiency in the absence of cooling is η_p, then it may be shown [5] that

$$T/p^\sigma = C/(1 + \psi)^\delta, \tag{4.33}$$

where $\sigma = (\gamma - 1)\eta_p/\gamma$ and $\delta = 1 - (x/\theta)$. At the exit state E,

$$T_E/T_1 = \theta/r^\sigma(1 + \psi_E)^\delta. \tag{4.34}$$

Alternatively,

$$T/p^{\sigma'} = \text{constant}, \tag{4.35}$$

where $\sigma' = (\gamma - 1)\eta_p/\gamma(1 - \lambda)$, and λ is obtained from heat transfer analysis as indicated earlier. A 'modified' polytropic efficiency is $\eta_p' = \eta_p/(1 - \lambda)$, so that $\sigma' = \eta_p'(\gamma - 1)/\gamma$. The turbine temperature at exit is then given by

$$T_E/T_1 = \theta/r^{\sigma'}. \tag{4.36}$$

Clearly, if λ is zero (no heat transfer), then the normal polytropic relation holds. A point of interest is that if $\eta_p = (1 - \lambda)$ then $\eta_p' = 1$ and the expansion becomes isentropic (but not reversible adiabatic).

4.2.2.5. *Comment*

For the various reversible cycles described in Section 4.2.1, the thermal efficiency was the same, independent of the number of cooling steps. This is not the case for the irreversible cycles described in this section. Both the thermal efficiency and the turbine exit temperature depend on the number and nature of cooling steps (whether the cooling air is throttled or not).

4.3. Open cooling of turbine blade rows—detailed fluid mechanics and thermodynamics

4.3.1. *Introduction*

The preliminary a/s analyses of turbine cooling described above contained two assumptions:
(i) open cooling with the cooling fraction known;
(ii) adiabatic mixing at constant pressure (low velocities were assumed, stagnation and static conditions being the same).

In Chapter 5 (and Appendix A), the detailed fluid mechanics and thermodynamics involved in cooling an individual turbine blade row are discussed, enabling ψ to be determined so that computer calculations for 'real' plants can be made. Here we continue to assume that the cooling fraction is known, but use a computer code based on real gas data to undertake parametric estimates of plant performance (the code developed by Young [11] was employed, as in Chapter 3 for uncooled cycles).

We concentrate here on open loop cooling in which compressor air mixes with the mainstream after cooling the blade row, the system most widely used in gas turbine plants (but note that a brief reference to closed loop steam cooling in combined cycles is made later, in Chapter 7). For a gas turbine blade row, such as the stationary entry nozzle guide vane row where most of the cooling is required, the approach first described here (called the 'simple' approach) involves the following:

(a) assuming a value of ψ, use of the steady flow energy equation to determine the overall change in the mainstream flow temperature from combustion temperature to rotor inlet temperature;

(b) determining the magnitude of the stagnation pressure drop involved in the process (which is also dependent on the magnitude of ψ).

From (a) and (b), the stagnation pressure and temperature can thus be calculated at exit from the cooled row; they can then be used to study the flow through the next (rotor) row. From there on a similar procedure may be followed (for a rotating row the relative $(T_0)_{rel}$ and $(p_0)_{rel}$ replace the absolute stagnation properties). In this way, the work output from the complete cooled turbine can be obtained for use within the cycle calculation, given the cooling quantities ψ.

Young and Wilcock [7] have recently provided an alternative to this simple approach. They also follow step (a), but rather than obtaining p_0 as in (b) they determine the constituent entropy increases (due to the various irreversible thermal and mixing effects). Essentially, they determine the downstream state from the properties T_0 and the entropy s, rather than T_0 and p_0. This approach is particularly convenient if the rational efficiency of the plant is sought. The lost work or the irreversibility $(\sum I = T_0 \sum \Delta S)$ may be subtracted from the ideal work $[-\Delta G_0]$ to obtain the actual work output and hence the rational efficiency,

$$(\eta)_R = 1 - T_0 \sum \Delta S / [-\Delta G_0]. \tag{4.37}$$

These two approaches may be shown to be thermodynamically equivalent and, given the same assumptions, will lead to identical results for the state downstream of a cooled row (if the input conditions are the same—see the published discussion of Ref. [7]). But the Young and Wilcock method gives a fuller understanding of the details of the cooling process.

Here we first describe the 'simple' approach, assuming that ψ is known, and describe how p_0 and T_0 downstream of the cooled row are obtained (steps (a) and (b) above). We then briefly describe the Young/Wilcock approach which leads to the determination and summation of the component entropy increases, again for a given ψ.

We defer to Chapter 5 (and Appendix A) a description of how the required cooling fraction ψ (and the heat transferred) can be obtained from heat transfer analysis, following the work of Holland and Thake [12].

4.3.2. *The simple approach*

Fig. 4.8 shows the open cooling process in a blade row diagrammatically. The heat transfer Q, between the hot mainstream (g) and the cooling air (c) inside the blades, takes place from control surface A to control surface B, i.e. from the mainstream (between combustion outlet state 3g and state Xg), to the coolant (between compressor outlet state 2c and state Xc). The injection and mixing processes occur within control surface C (between states Xg and Xc and a common fully mixed state 5m, the rotor inlet state). The flows through A plus B and C are adiabatic in the sense that no heat is lost to the environment outside these control surfaces; thus the entire process (A + B + C) is adiabatic. We wish to determine the mixed out conditions downstream at station 5m.

4.3.2.1. *Change in stagnation enthalpy (or temperature) through an open cooled blade row*

The total enthalpy change across the whole (stationary) cooled blade row is straightforward and is obtained for the overall process (i.e. the complete adiabatic flow through control surfaces (A + B) plus (C)). Even though there is a heat transfer Q 'internally' between the unit mainstream flow and the cooling air flow ψ, from A to B, the overall process is adiabatic.

In the simplified a/s analysis of Section 4.2 we assumed identical and constant specific heats for the two streams. Now we assume semi-perfect gases with specific heats as functions of temperature; but we must also allow for the difference in gas properties between the cooling air and the mainstream gas (combustion products). Between entry states (mainstream gas 3g, and cooling air, 2c) and exit state 5m (mixed out), the steady flow energy equation, for the flow through control surfaces (A + B) and C, yields, for a stationary blade row,

$$(h_0)_{3g} + \psi(h_0)_{2c} = (1 + \psi)(h_0)_{5m}. \tag{4.38}$$

It is assumed that the entry gas (g), the cooling air (c) and the mixed exit gas (m) are all semi-perfect gases with enthalpies measured from the same temperature datum (absolute temperature, $T = 0$). The specific heat at constant pressure of the mixture in state 5m

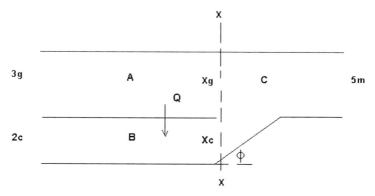

Fig. 4.8. Mixing of cooling air with mainstream flow.

is given by

$$(c_p)_{5m} = [(c_p)_{5g} + \psi(c_p)_{5c}]/(1 + \psi), \qquad (4.39)$$

and hence

$$c_{pg}[(T_0)_{3g} - (T_0)_{5g}] = \psi c_{pc}[(T_0)_{5c} - (T_0)_{2c}], \qquad (4.40)$$

where the specific heats are now mean values over the relevant temperature range.

These equations enable the exit temperature T_{05m} to be determined. Alternatively, the exit enthalpy can be obtained directly from

$$(h_0)_{3g} - (h_0)_{5g} = \psi[(h_0)_{5c} - (h_0)_{2c}], \qquad (4.41)$$

if tables of gas properties are used instead of specific heat data.

4.3.2.2. *Change of total pressure through an open cooled blade row*

It has already been shown that (stagnation) pressure losses have an appreciable effect on cycle efficiency (see Section 3.3), so as well as obtaining the enthalpy change, it is important to determine the stagnation pressure change in the whole cooling process.

To determine the overall change in total pressure we must now consider the three control surfaces A, B and C of Fig. 4.8 separately.

For the fluid streams flowing through control surface A and B we may regard each as undergoing a Rayleigh process—a compressible fluid flow with friction and heat transfer. According to Shapiro [13], in such a process the change in total pressure Δp_0 over a length dx is related to the change in stagnation temperature ΔT_0 and to the skin friction as

$$\Delta p_0/p_0 = -(\gamma M^2/2)[(\Delta T_0/T_0) - (4f\,dx/d_h)], \qquad (4.42)$$

where M is the Mach number, f the skin friction coefficient and d_h the hydraulic mean diameter of the duct. For the mainstream gas flow in control surface A, $(\Delta T_0)_g = -Q/c_{pg}$; and for the cooling air flow in B, $(\Delta T_0)_c = +Q/\psi c_{pc}$, where Q is the heat transferred, which is determined from heat transfer analysis as described in Chapter 5 and Appendix A.

In the simple approach, the change p_0 due to Q (the first term in Eq. (4.42)) is usually ignored for both streams. The change of p_0 due to frictional effects in the mainstream flow is usually included in the basic polytropic efficiency (η_p) of the uncooled flow, so that

$$[(p_0)_{3g} - (p_0)_{Xg}]/(p_0)_{3g} \approx \gamma M_{3g}^2[1 - \eta_p]/2 \qquad (4.43)$$

is already known. The change of p_0 due to friction in the coolant flow through the complex internal geometry is usually obtained using an empirical friction factor k so that

$$[(p_0)_{2c} - (p_0)_{Xc}]/(p_0)_{2c} = k(M_{2c})^2/2. \qquad (4.44)$$

Thus, p_0 and T_0 at exit from the control surfaces A and B are given by
A (mainstream gas)

$$(T_0)_{Xg} = (T_0)_{3g} - Q/c_{pg}, \qquad (p_0)_{Xg} \approx (p_0)_{3g}\{1 - \gamma M_{3g}^2[1 - \eta_p]/2\}, \qquad (4.45)$$

B (coolant air)

$$(T_0)_{Xc} = (T_0)_{2c} + Q/\psi c_{pc}, \qquad (p_0)_{Xc} \approx (p_0)_{2c}(1 - kM_{2c}^2/2). \qquad (4.46)$$

We can then proceed to determine the changes across control surface C. The final total temperature $(T_0)_{5m}$ has already been obtained but the total pressure $(p_0)_{2m}$ has to be determined. An expression given by Hartsel [14] for the mainstream total pressure loss in this adiabatic mixing process again goes back to the simple one-dimensional momentum analysis given by Shapiro [13] for the flow through control surface C illustrated in Fig. 4.8. Hartsel developed Shapiro's table of influence coefficients to allow for a difference between the total temperature of the injected flow (now $(T_0)_{Xc}$) and the mainstream $(T_0)_{Xg}$):

$$\Delta p_0/p_0 = [1 - (p_0)_{5m}/(p_0)_{Xg}]$$

$$= -(\psi\gamma M_{Xg}^2/2)\{1 + [(T_0)_{Xc}/(T_0)_{Xg}] - 2y \cos \phi\}. \qquad (4.47)$$

Here y is the ratio of the velocity of the injected coolant to that of the free stream $(y = V_c/V_g)$, M_{Xg} the Mach number of the free stream and ϕ the angle at which the cooling air enters the mainstream (Fig. 4.8).

The value of y has to be determined; an approximation suggested by Hartsel is to take $(p_0)_{Xc} = (p_0)_{Xg}$, so that $V_c/V_g \approx [(T_0)_{Xc}/(T_0)_{Xg}]^{1/2}$, since the static pressures must be the same where the coolant enters. A sufficient approximation might be to take $(T_0)_{Xg}$ as the exit temperature from the combustion chamber and $(T_0)_{Xc}$ as the exit temperature from the compressor (i.e. again ignoring Q in Eqs. (4.45) and (4.46)).

A more sophisticated approach would not only take account of Eqs. (4.45) and (4.46) to give the two stagnation temperatures at exit from control surfaces A and B, but it would also not assume the total pressures of coolant and mainstream to be the same. For the first nozzle guide vane row these can be derived by accounting for losses as follows:

(i) in the *mainstream* (g), the stagnation pressure at delivery from the compressor less Δp_{0CC} in the combustion process, and Δp_0 in the nozzle row itself (as in control surface A, due to friction and the heat transfer away from the mainstream gas if included);

(ii) in the *coolant* air stream (c), the stagnation pressure at extraction from the compressor less a loss Δp_{0D} (in the ducting and disks before coolant enters the blade itself), and Δp_{0B} (in the blading heat transfer process in control surface B due to both friction and heat transfer, if included).

The total pressures at X may thus be determined, as $(p_0)_{Xg}$ and $(p_0)_{Xc}$. If, as Hartsel implies, the mainstream Mach number at X (M_{Xg}) is also known, which means that the static pressure at the mixing plane (p_X) is also known, M_{Xc} may also be determined from $(p_0)_{Xc}$. The two different velocities V_c and V_g are then obtained, together with the required value of y for Eq. (4.47).

But there is a further subtle point here in determining y, as implied by Young and Wilcock. With $[(p_0)_{Xc}/p_X]$ known, not only is the Mach number M_{Xc} known but also the non-dimensional mass flow, $\{\psi[R(T_0)_{Xc}]^{1/2}/A_{Xc}(p_0)_{Xc}\}$, may be obtained. This means that the area A_{Xc}, required to pass the coolant flow, is also determined. Obviously a degree of successive approximation should be involved in obtaining the full solution to the complete cooling flow process.

An empirical development of the approach described above uses experimental cascade data, obtained with and without coolant discharge, to obtain an overall relationship between the total cooling flow through the blade row (ψ) and the *extra* stagnation pressure loss arising from injection of the cooling air. In film cooling, the air flow leaves the blade surface at various points round the blade profile causing variable loss (noting that injection near the trailing edge causes little total pressure loss—it may even reduce the basic loss in the wake). If there is an elementary amount of air $d\psi$ at a particular location where the injection angle is ϕ, then an overall figure for the extra total pressure loss due to coolant injection in a typical blade row can be obtained by 'integrating' the Hartsel equation (4.47) round the blade profile [3]. An overall exchange factor for the *extra* blade row stagnation pressure mixing loss in the row can thus be obtained in the form

$$\Delta p_0/p_0 = -\kappa\psi, \tag{4.48}$$

to be used in the subsequent cycle calculations. Alternatively, Eq. (4.48) can be converted into a modified small stage or polytropic efficiency, $\eta_p \approx \eta_{stage}$

$$\Delta\eta_{stage}/\eta_{stage} = \kappa'\psi, \tag{4.49}$$

using the relationship given in Ref. [3],

$$\kappa/\kappa' = [\Delta\eta_{stage}/\eta_{stage}]/\left[\sum\Delta p_t/p_t\right] \approx [(\gamma-1)/\gamma](x_{stage}-1), \tag{4.50}$$

in which $x_{stage} = r_{stage}^{(\gamma-1)/\gamma}$.

4.3.3. Breakdown of losses in the cooling process

The simple approach described before involves approximations, particularly to obtain the stagnation pressure loss. The full determination of $(p_0)_{5m}$ and $(T_0)_{5m}$ from the various equations given above can lead to an approximation for the downstream entropy (s_{5m}), using the Gibbs relation applied between stagnation states,

$$T_0\Delta s = \Delta h_0 - \Delta p_0/\rho_0. \tag{4.51}$$

If the outlet specific entropy s_{5m} is determined in this way the gross entropy generation in the whole process is also obtained,

$$\Delta S = (1+\psi)s_{5m} - (s_{3g} + \psi s_{2c}), \tag{4.52}$$

and hence the total irreversibility $I = T_0\Delta S$. However, this does not give details on how the various irreversibilities arise in the cooling process.

Young and Wilcock [7] provided a much more rigorous approach which includes an illuminating discussion of how the losses arise in the cooling process. They prefer to address the problem by breaking the overall flow into flows through the 'component' control surfaces of Fig. 4.8 and determining the various entropy changes directly. Their breakdown of the gross entropy then involves writing

$$\Delta S = \Delta S_{external} + \Delta S_{metal} + \Delta S_{internal}. \tag{4.53}$$

Here $\Delta S_{internal}$ is the entropy increase of the cooling fluid in control surface B due to friction and the heat transfer (Q, in), ΔS_{metal} is the entropy created in the metal between the mainstream and the coolant (or metal plus thermal barrier coating if present) due to temperature difference across it, $\Delta S_{external}$ is the entropy increase in the mainstream flow within control surface A before mixing due to heat transfer (Q, out), plus the various entropy increases due to the mixing process itself in control surface C.

The reader is referred to the original papers for detailed analysis, where the various components of entropy generation and irreversibility are defined. The advantage of this work is not only that it involves less approximation but also that it is revealing in terms of the basic thermodynamics. It should also be used by designers who should be able to see how design changes relate to increased or decreased local loss.

4.4. Cycle calculations with turbine cooling

In order to make a preliminary assessment of the importance of turbine cooling in cycle analysis, the real gas calculations of a simple open uncooled cycle, carried out in Chapter 3 for various pressure ratios and combustion temperatures, are now repeated with single step turbine cooling, i.e. including cooling of the first turbine row, the stationary nozzle guide vanes.

Here the magnitudes of the cooling flow fractions are assumed, together with the extra stagnation pressure loss due to mixing. Subsequently, in Chapter 5, the calculations are repeated for cooling flow fractions accurately assessed from heat transfer analysis, together with associated total pressure losses. But the present investigation concentrates on whether the conclusion derived from the a/s analyses—that cooling makes relatively little difference to plant thermal efficiency—remains valid when real gas effects are included.

For the purpose of the current calculations the cooling flow fractions were assumed to increase linearly with combustion temperature, from 0.05 at 1200°C. Thus, the following values of cooling fraction were assumed: 0.05 at 1200°C; 0.075 at 1400°C; 0.10 at 1600°C; 0.125 at 1800°C; 0.15 at 2000°C.

The choice of these values is arbitrary. In practice, the cooling fraction will depend not only on the combustion temperature but also on the compressor delivery temperature (i.e. the pressure ratio), the allowable metal temperature and other factors, as described in Chapter 5. But with ψ assumed for the first nozzle guide vane row, together with the extra total pressure loss involved ($\kappa = 0.07$ in Eq. (4.48)), the rotor inlet temperature may be determined. These assumptions were used as input to the code developed by Young [11] for cycle calculations, which considers the real gas properties.

Fig. 4.9 shows the results of calculations based on these assumptions in comparison with the uncooled calculations (the other assumptions were those listed for the earlier uncooled calculations in Section 3.4.1). The (arbitrary) overall efficiency is shown plotted

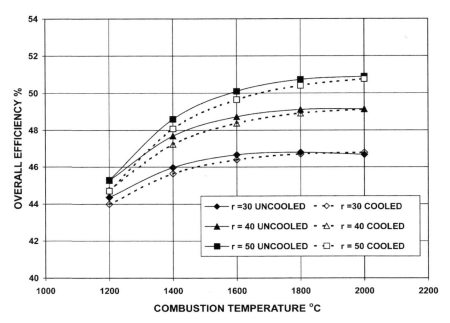

Fig. 4.9. Calculation of efficiency of simple [CBT] plants—single-step cooled [CBT]$_{IC1}$ and uncooled [CBT]$_{IU}$—as a function of maximum temperature (T_{cot}) with pressure ratio (r) as a parameter.

against combustion outlet temperature ($T_{cot} = T_3$) for various selected pressure ratios ($r = 30, 40, 50$). It is indeed clear that the drop in efficiency produced by turbine cooling is small, as anticipated in the a/s analyses developed earlier in this chapter. This drop decreases with increasing combustion temperature as anticipated in the a/s analysis leading to Eq. (4.24); indeed at the highest combustion temperatures there appears to be no drop in thermal efficiency at all. It is explained later in Chapter 5 that this is a small real gas effect brought about by the change in the constitution of the combustion products, and in particular the dominant effect of the water vapour content on the mean specific heat.

Fig. 4.10 shows more fully calculated overall efficiencies (for turbine cooling only) replotted against isentropic temperature ratio for various selected values of $T_3 = T_{cot}$. This figure may be compared directly with Fig. 3.9 (the a/s calculations for the corresponding CHT cycle) and Fig. 3.13 (the 'real gas' calculations of efficiency for the uncoooled CBT cycle). The optimum pressure ratio for maximum efficiency again increases with maximum cycle temperature T_3.

The (arbitrary) overall efficiency and specific work quantities obtained from these calculations are illustrated as carpet plots in Fig. 4.11. It is seen that the specific work is reduced by the turbine cooling, which leads to a drop in the rotor inlet temperature and the turbine work output. Again this conclusion is consistent with the preliminary analysis and calculations made earlier in this chapter.

A final calculation illustrates the earlier discussion on the difference between combustion temperature $T_{cot} = T_3$ and rotor inlet temperature $T_{rit} = T_5$. Fig. 4.12 shows

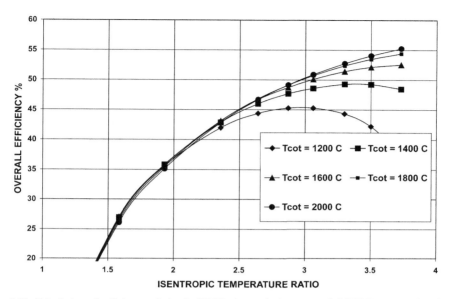

Fig. 4.10. Calculation of efficiency of simple [CBT] plant—single-step cooled [CBT]$_{IC1}$ as a function of isentropic temperature ratio with maximum temperature (T_{cot}) as a parameter.

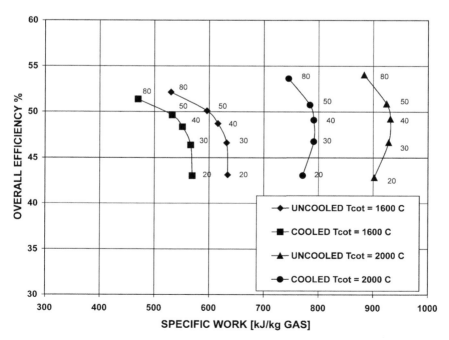

Fig. 4.11. Calculation of efficiency of simple [CBT] plants—single-step cooled [CBT]$_{IC1}$ and uncooled [CBT]$_{IU}$—as a function of specific work with pressure ratio (r) and maximum temperature (T_{cot}) as parameters and with $\eta_{pC} = \eta_{pT} = 0.9$, $T_{bl} = 1073$ K (after Ref. [5]).

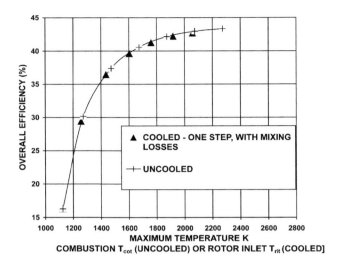

Fig. 4.12. Calculation of efficiency of [CBT] plant: uncooled [CBT]$_{IU}$ as a function of combustion temperature (T_{cot}); single-step cooled [CBT]$_{IC1}$ as a function of rotor inlet temperature (T_{rit}). Pressure ratio $r = 30$, $\eta_C = 0.8$, $\eta_T = 0.9$, $T_{bl} = 1123$ K (after Ref. [5]).

a single step cooling calculation of overall efficiency (for a pressure ratio of 20) plotted against both T_3 and T_5. It is seen that data expressed as $\eta_O(T_5)$ does in fact almost fall on the uncooled efficiency line $\eta_O(T_3)$, the effect anticipated in Section 4.2.2.2, where a/s analysis was used leading to the diagram of Fig. 4.6c.

4.5. Conclusions

It has been shown from *air-standard* cycle analysis that
(a) plant efficiency drops relatively little due to turbine cooling;
(b) the efficiency of the cooled turbine plant, when expressed as a function of the rotor inlet temperature (T_5), is virtually identical to the efficiency of the uncooled plant when expressed as a function of combustion temperature (T_3);
(c) the difference between uncooled and cooled efficiency decreases at high combustion temperature;
(d) the rate of increase of thermal efficiency of the cooled cycle falls with increasing T_3; but there is no prediction of a maximum efficiency being attained at high T_3.

These conclusions are broadly confirmed by *real gas* calculations for single step cooling with arbitrary assumptions for cooling flow fractions.

But it appears that thermal efficiency does tend towards a maximum level with increasing combustion temperature. More realistic calculations of highly cooled turbines are given in the next chapter, after a brief description of the heat transfer analysis involved in the determination of cooling flow quantities.

References

[1] Mukherjee, D.K. (1976), Design of turbines, using distributed or average losses; effect of blading, AGARD 195, 8-1–8-13.

[2] Chiesa, P., Consonni, S., Lozza, G. and Macchi, E. (1993), Predicting the ultimate performance of advanced power cycles based on very high temperatures, ASME paper 93-GT-223.

[3] Horlock, J.H., Watson, D.E. and Jones, T.V. (2001), Limitations on gas turbine performance imposed by large turbine cooling flows, ASME J. Engng Gas Turbines Power 123(3), 487–494.

[4] Hawthorne, W.R. and Davis, G.de V. (1956), Calculating gas turbine performance, Engineering 181, 361–367.

[5] Horlock, J.H. (2001), Basic thermodynamics of turbine cooling, ASME J. Turbomachinery 123(3), 583–592.

[6] Denton, J.D (1993), Loss mechanisms in turbomachines, ASME paper 93-GT-435.

[7] Young, J.B. and Wilcock, R.C. (2002), Modelling the air-cooled gas turbine, Part1, ASME J. Turbomachinery 124, 207–213.

[8] Traupel, W. (1966), Thermische Turbomaschinen, Springer Verlag, Berlin.

[9] Hawthorne, W.R. (1956), The thermodynamics of cooled turbines, Parts I and II, Proc. ASME 78, 1765 (see also p. 1781).

[10] El-Masri, M.A. (1987), Exergy analysis of combined cycles. Part 1. Air-cooled Brayton-cycle gas turbines, ASME J. Engng Power Gas Turbines 109, 228–235.

[11] Young, J.B. (1998), Computer-based project on combined-cycle power generation, Cambridge University Internal Report.

[12] Holland, M.J. and Thake, T.F. (1980), Rotor blade cooling in high pressure turbines, AIAA J. Aircraft 17(6), 412–418.

[13] Shapiro, A.H. (1953), The dynamics and thermodynamics of compressible fluid flow, Ronald Press, New York.

[14] Hartsel, J.E. (1972), Prediction of effects of mass-transfer cooling on the blade-row efficiency of turbine airfoils, AIAA paper 72-11.

Chapter 5

FULL CALCULATIONS OF PLANT EFFICIENCY

5.1. Introduction

In Chapter 4 calculations were made on the overall efficiency of CBT plants with turbine cooling, the fraction of cooling air (ψ) being assumed arbitrarily. In this chapter, we outline more realistic calculations, with the cooling air fraction ψ being estimated from heat transfer analysis and experiments.

There are several papers in the literature which give details of cycle calculations, and include details of how the cooling flow quantity may be estimated and used. Here we describe one such approach used by the author and his colleagues. Initially, we summarise how ψ can be obtained (fuller details are given in Appendix A). We then illustrate how this information is used in calculations, once again using a computer code in which real gas effects are included.

Subsequently, we refer briefly to other comparable studies, including the calculations of exergy losses and rational efficiency. Finally, we show the 'real gas' exergy calculations for two practical plants—[CBT]$_{\mathrm{I}}$ and [CBTX]$_{\mathrm{I}}$.

5.2. Cooling flow requirements

The method devised by Holland and Thake [1] for estimating the cooling air (w_c), as a fraction of mainstream entry flow to a blade row (w_g), i.e. $\psi = w_c/w_g$, was described by Horlock et al. [2] and is reproduced in Appendix A; Fig. A.1 shows diagrammatically the notation employed there and the same symbols are defined and used below.

5.2.1. Convective cooling

Consider first a convectively cooled blade row (Fig. A.1a). It is shown in Appendix A that the mass flow of cooling air (w_c) required for a mass flow of mainstream gas (w_g), entering at temperature T_{gi}, is given by

$$\psi = w_c/w_g = Cw^+, \tag{5.1}$$

where w^+ is a 'temperature difference ratio' defined as

$$w^+ = (T_{\mathrm{gi}} - T_{\mathrm{bl}})/(T_{\mathrm{bl}} - T_{\mathrm{ci}}), \tag{5.2}$$

with T_{bl}, the allowable blade temperature and T_{ci}, the cooling air entry temperature.

If ε_0 is the blade cooling effectiveness, defined as

$$\varepsilon_0 = (T_{gi} - T_{bl})/(T_{gi} - T_{ci}), \tag{5.3}$$

and η_{cool} is the cooling efficiency,

$$\eta_{cool} = (T_{co} - T_{ci})/(T_{bl} - T_{ci}), \tag{5.4}$$

in which T_{co} is the cooling air outlet temperature before mixing, then it follows that

$$w^+ = C\varepsilon_0/\eta_{cool}(1 - \varepsilon_0). \tag{5.2a}$$

The 'constant' C is

$$C = St_g(A_{gs}/A_{gx})(c_{pg}/c_{pc}), \tag{5.5}$$

in which St_g is the *external* gas Stanton number, A_{gs} and A_{gx} are the gas surface and cross-sectional flow areas, and c_{pg}, c_{pc} are the gas and cooling air specific heats, respectively.

The cooling efficiency can be determined from the internal heat transfer. If T_{bl} is considered to be more or less constant, then it may be shown that

$$\eta_{cool} = 1 - \exp(-\xi), \tag{5.6}$$

where $\xi = (h_c A_{cs}/w_c c_{pc}) = (St_c A_{cs}/A_{cx})$, and St_c is now the internal cooling air Stanton number, A_{cs} and A_{cx} referring to surface and cross-sectional areas of the internal cooling air flow, respectively.

Experience gives values of ξ for various geometries, but St_c is found to be a weak function of Reynolds number, so in practice there is relatively little variation in cooling efficiency ($0.6 < \eta_{cool} < 0.8$). Thus, both C and η_{cool} do not vary greatly and if they are amalgamated into a single constant, $K = C/\eta_{cool}$, then

$$\psi = K\varepsilon_0/(1 - \varepsilon_0), \tag{5.7}$$

or

$$\psi = K(T_{gi} - T_{bl})/(T_{bl} - T_{ci}), \tag{5.8}$$

a form used by El-Masri [3] for his cycle calculations, and also used in the last chapter to relate T_{cot} and T_{rit}.

5.2.2. Film cooling

For a film cooled blade row, cooling air at outlet temperature T_{co} is discharged into the mainstream through the holes in the blade surface to form a cooling film (Fig. A.1b).

A film cooling effectiveness is now defined as

$$\varepsilon_F = (T_{gi} - T_{aw})/(T_{gi} - T_{co}), \tag{5.9}$$

where T_{aw} is the adiabatic wall temperature.

A new 'temperature difference ratio' W^+ is written as

$$W^+ = [T_{aw} - T_{bl}]/[T_{co} - T_{ci}]$$

$$= [\varepsilon_0 - (1 - \eta_{cool})\varepsilon_F - \varepsilon_0\varepsilon_F\eta_{cool}]/\eta_{cool}(1 - \varepsilon_0), \qquad (5.10)$$

and it is shown in Appendix A that the cooling fraction is now given by

$$\psi = (w_c/w_g) = (c_{pg}/c_{pc})(A_{gs}S_{tg}/A_{gx})\mu W^+, \qquad (5.11)$$

where

$$\mu = h_{fg}/[h_g(1 + B)], \qquad (5.12)$$

in which (h_{fg}/h_g) is the ratio of the heat transfer coefficient under film cooling conditions (h_{fg}) to the convectively cooled heat transfer coefficient (h_g), and $B = h_{fg}t/k$ is the Biot number, which takes account of a thermal barrier coating (TBC) of thickness t and conductivity k. In practice, h_{fg} increases above h_g, and $(1 + B)$ is increased as TBC is added. For the purposes of the cycle calculations described below, μ is taken as unity so that

$$\psi = CW^+, \qquad (5.13)$$

where C is the same constant as the one for convective cooling only.

5.2.3. *Assumptions for cycle calculations*

In the cycle calculations described below [2], film cooling was assumed. Further, as described in Appendix A, various assumptions were made for the critical constants, as follows. The constant C in Eq. (5.13) was taken as 0.045, and within W^+, the cooling efficiency η_{cool} as 0.7 and the film cooling effectiveness ε_F as 0.4. All were assumed to be constant over the range of cooling flows considered.

In a particular blade row, for a given gas entry temperature T_{gi}, a cooling air entry temperature T_{ci}, and an assumed allowable blade metal temperature T_{bl}, the blade cooling effectiveness ε_0 is obtained. With $\varepsilon_F = 0.4$ and $\eta_{cool} = 0.7$, W^+ then follows from Eq. (5.10). With $C = 0.045$ the cooling air flow fraction ψ is obtained from Eq. (5.13).

5.3. Estimates of cooling flow fraction

The results of calculations for the cooling air flow fractions in the first (nozzle guide vane) row of the turbine, based on the assumptions outlined in Section 5.2 for film cooled blading, are illustrated in Fig. 5.1. The entry gas temperature T_{gi} was taken as the combustion temperature $T_{cot} = T_3$ and the cooling air temperature as the compressor delivery temperature T_2. The cooling air required is shown here as a fraction of the exhaust gas flow, i.e. as $\psi/(1 + \psi)$, plotted against compressor pressure ratio and combustion temperature for an allowable blade metal temperature, $T_{bl} = 800°C$. Also shown are

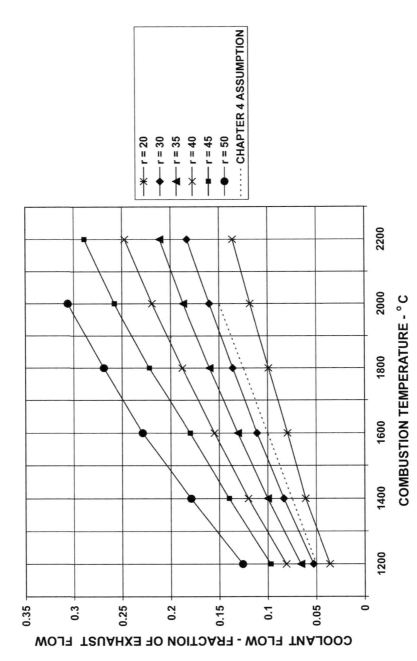

Fig. 5.1. Calculated coolant air fractions for single step cooling (of nozzle guide vanes), as a function of combustion temperature with pressure ratio as a parameter.

the arbitrary assumptions made in Chapter 4 for the calculations to illustrate the changes in thermal efficiency for gas turbine plants in which single-step cooling is introduced.

The cooling fraction obviously increases with combustion temperature, but the compressor pressure ratio (and hence the cooling air temperature T_2) is also critically important. It is seen that the arbitrary assumptions made for ψ in Chapter 4 (linearly increasing with the combustion temperature $T_{cot} = T_3$) would be approximately valid for a cycle with a pressure ratio just below 30.

5.4. Single step cooling

The results of a set of computer calculations for a CBT plant with single-step cooling (i.e. of the first stage nozzle guide vanes) are illustrated in Fig. 5.2, in the form of (arbitrary) overall thermal efficiency (η_O) against pressure ratio (r) with the combustion temperature T_{cot} as a parameter, and in Fig. 5.3 as η_O against T_{cot} with r as a parameter.

Young's computer code [4] was used for these efficiency calculations. It involves an assumption that the mainstream gas is expanded through a nominal (small) pressure ratio, mixed with cooling air at compressor delivery conditions and this mixed gas then expanded through the full turbine pressure ratio. Within the calculations, the values of ψ given in Fig. 5.1 were also used to derive the extra stagnation pressure loss associated with mixing (as described in Section 4.3.2.2 leading to Eq. (4.47), with the empirical constant κ taken as 0.07). This extra stagnation pressure loss was added to the assumed stagnation pressure loss in combustion, $(\Delta p_0/p_0)_{CC} = 0.03$.

Fig. 5.2 shows that for the single-step cooled CBT plant at a given combustion temperature, the overall efficiency of the cooled gas turbine efficiency increases with pressure ratio initially but, compared with an uncooled cycle, reaches a maximum at a lower optimum pressure ratio. Fig. 5.3 shows that for a given pressure ratio the efficiency generally increases with the combustion temperature T_{cot} even though the required cooling fraction increases.

Fig. 5.4 shows a carpet plot of overall efficiency against specific work for the cooled [CBT]$_{IC1}$ plant (single step) with pressure ratio and combustion temperature as parameters. As shown earlier, by the preliminary air standard analysis and the subsequent calculations in Chapter 4, there are relatively minor changes of thermal efficiency compared with the uncooled plant [CBT]$_{IUC}$, but there is a major effect in the reduction of specific work.

5.5. Multi-stage cooling

At very high combustion temperatures, it is not sufficient that the first blade row alone needs to be cooled. In practice, up to half a dozen rows may be cooled in an industrial gas turbine, if the combustion temperature is high and the allowable blade metal temperature is low. The cooling fractions for each of the cooled rows must be estimated and used in the cycle calculations, which now become complex.

Illustrations of such calculations, for an open cycle [CBT]$_{IC3}$ plant, were given by Horlock et al. [2], in which it was assumed that three blade rows were film cooled, the two

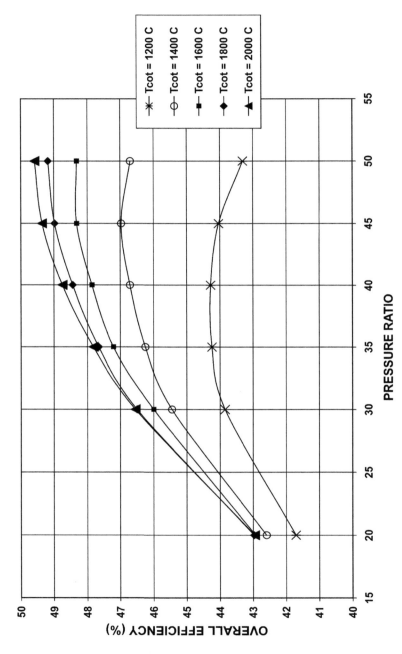

Fig. 5.2. Overall efficiency of [CBT]$_{IC1}$ plant with single-step cooling of NGVs, as a function of pressure ratio with combustion temperature as a parameter.

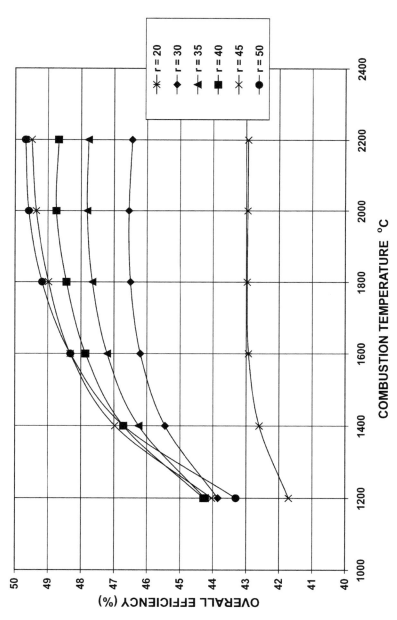

Fig. 5.3. Overall efficiency of [CBT]$_{IC1}$ plant with single-step cooling of NGVs, as a function of combustion temperature with pressure ratio as a parameter.

Advanced gas turbine cycles

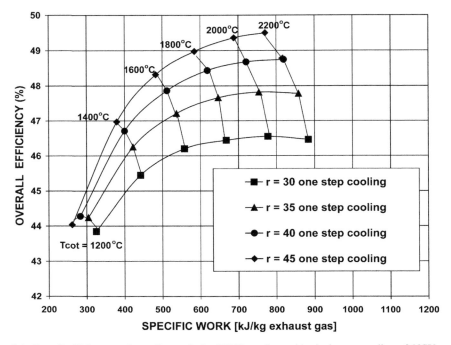

Fig. 5.4. Overall efficiency and specific work for [CBT]$_{IC1}$ plant with single-step cooling of NGVs, with combustion temperature and pressure ratio as parameters (after Ref. [5], Chapter 4).

rows of the first turbine stage and the stationary nozzle guide vanes of the second stage. As in the single-step cooling calculations described before, film cooling was assumed and the Holland and Thake approach was followed to determine the cooling air required in each of these blade rows.

From the combustion temperature T_{cot} and an assumed first stage pressure ratio (3:1), the 'mixed out' gas temperature at exit from the first stage (T_{E1}) was obtained and this was taken as the gas entry temperature for the second stage (third blade row). The entry (relative) stagnation temperature for the first stage rotor (the second turbine blade row) was obtained by interpolation between T_{E1} and T_{cot}, assuming 50% reaction in the first stage. The cooling air inlet temperature was taken as the compressor delivery temperature, $T_{ci} = T_2$ for all three rows. This would have led to the estimation of coolant flow in the second and third rows being somewhat more than needed as the cooling air could theoretically be tapped at a lower pressure (and therefore lower cooling temperature). But in practice the pressure loss through the supply ducts and past the turbine disks can be substantial and compressor delivery pressure may have to be used anyway. The cooling fractions thus obtained for the three rows are shown in Fig. 5.5; obviously the first row requires most cooling, the fractions for the subsequent rows decrease and it is assumed that the fourth row requires no cooling.

The cycle calculations for this multi-cooling then proceeded in a similar fashion to those for the single-step cooling calculations of Section 5.4 (full details are given in Ref. [2]).

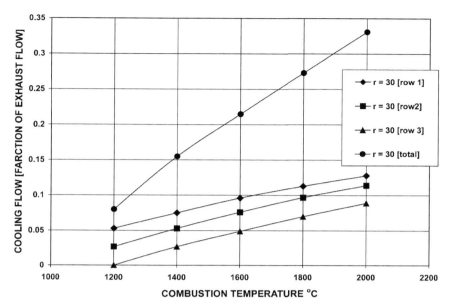

Fig. 5.5. Calculated coolant air fractions for three step cooling (of first stage and second rotor row).

Fig. 5.6 shows the results of a set of computer calculations for the $[CBT]_{IC3}$ plant in the form of (arbitrary) overall efficiency (η_O) against pressure ratio (r) with the combustion temperature T_{cot} as a parameter. Fig. 5.7 shows η_O plotted against T_{cot} with r as a parameter and Fig. 5.8 shows a contour plot of η_O against T_{cot} and r. There is a flat efficiency plateau around $T_{cot} \approx 1750°C$, less than the maximum value used in these calculations, which approaches the stoichiometric limit.

The changes in the form of these graphs for three step cooling, compared with those for single-step cooling (Figs. 5.2 and 5.3), are most significant. They indicate that the overall efficiency of such a CBT plant may reach a limiting value, just over 44% at $T_{cot} \approx 1750°C$ and $r = 35$ for the assumptions made here ($\eta_p = 0.9$, $(\Delta p_0)_{CC} = 0.03$, with three rows of cooling each with compressor delivery air); whereas for single-step cooling the incentive is to keep raising T_{cot} together with the corresponding pressure ratio. But it should be emphasised that this conclusion is much dependent on the estimates for cooling flow fractions.

Fig. 5.9 shows a carpet plot of thermal efficiency for three step cooling. Now the picture is different from the corresponding carpet plot of Fig. 5.4 for single stage cooling, with the overall efficiencies collapsing into a narrow band around 44%, for temperatures T_{cot} between 1600 and 2000°C and for pressure ratios 30, 35 and 40. Advantages in thermal efficiency for both uncooled and single step cooling (at high T_{cot} and high pressure ratio) are now negated because of the large cooling flows required for three step cooling. However, the higher combustion temperature continues to give advantage in the larger specific work.

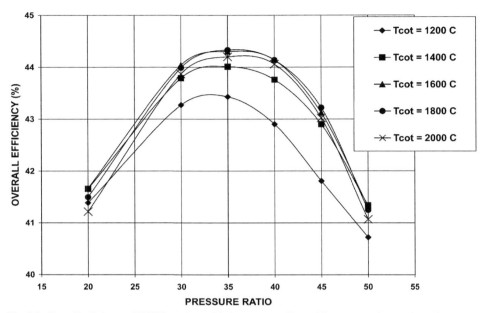

Fig. 5.6. Overall efficiency of [CBT]$_{IC3}$ plant with three step cooling (of first stage and second nozzle row) as a function of pressure ratio with combustion temperature as a parameter (after Ref. [2]).

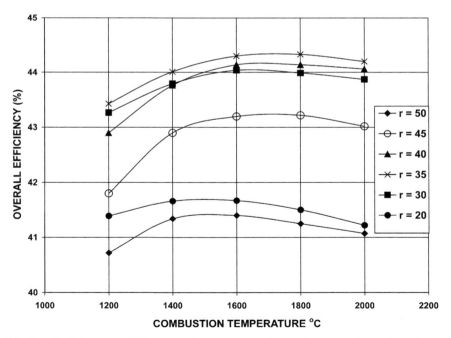

Fig. 5.7. Overall efficiency of [CBT]$_{IC3}$ plant with three step cooling (of first stage and second nozzle row) as a function of combustion temperature with pressure ratio as a parameter.

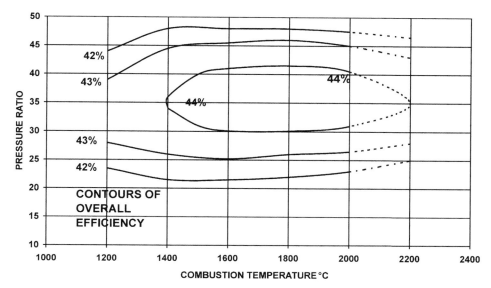

Fig. 5.8. Contours of overall efficiency for [CBT]$_{IC3}$ plant with three step cooling, against combustion temperature and pressure ratio.

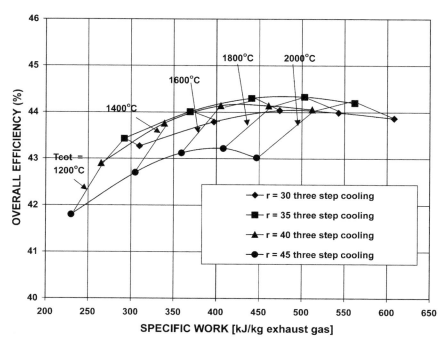

Fig. 5.9. Overall efficiency and specific work for [CBT]$_{IC3}$ plant with three step cooling (of first stage and second nozzle row), with combustion temperature and pressure ratio as parameters (after Ref. [5], Chapter 4).

5.6. A note on real gas effects

The real gas calculations with cooling as described above give indications of maxima in the plots of thermal efficiency against $T_3 = T_{cot}$ for a given pressure ratio (e.g. Fig. 5.3). These do not appear in air standard analysis such as that described in Chapter 3. The calculations of Chapter 4 showed that such maxima can occur not only for cooled but also, surprisingly, for uncooled calculations. Fig. 4.9 showed such graphs of η_O against T_{cot} to be very flat, but there was clearly a real gas effect independent of cooling at high T_{cot}. Recent detailed investigations of these real gas effects by Wilcock et al. [3] have revealed that this 'turnover effect' on uncooled efficiency at high values of T_{cot} is related to the changes in real gas properties (c_{pg} and γ_g) with both temperature and composition.

5.7. Other studies of gas turbine plants with turbine cooling

There are several studies in the literature which parallel the approach of Horlock et al. [2] described above. Some of the more important are listed here and briefly discussed.

Perhaps the most comprehensive set of papers were those by El-Masri and his colleagues in a series of publications in the 1980s. El-Masri describes his methods of predicting cooling flow requirements in Ref. [4] for combined convection and film cooling, and in Ref. [5] with thermal barrier coatings. The approach is similar but not identical to that described above. Following initial cycle calculations with working fluids with constant properties [6,7] El-Masri developed a computer code—GASCAN [8]—embracing real gas properties and used this in the second law calculations of air-cooled Brayton gas turbine cycles [9] and combined cycles [10]. These calculations presented details of exergy losses, work output and rational efficiency and gave some indication of an optimum combustion temperature yielding maximum efficiency (for a given pressure ratio), along the lines already described in this chapter.

Similarly, comprehensive calculations including turbine cooling were made by Lozza and his colleagues [11]. These calculations give results broadly similar to those described in this chapter but an important feature of this work involved a degree of parameterisation of the cooling methods—e.g. variation of the allowable blade temperature.

A third set of similar but simpler calculations were described by MacArthur [12] who applied aero-engine cooling technology to obtain improved performance of industrial type gas turbine power plants.

5.8. Exergy calculations

Once the state points are known round a cycle in a computer calculation of performance, the local values of availability and/or exergy may be obtained. The procedure for estimating exergy losses or irreversibilities was outlined in Chapter 2. Here we show such calculations made by Manfrida et al. [13] which were also presented in Ref. [14].

Fig. 5.10 shows the exergy losses as a fraction of the fuel exergy (including the partial pressure terms referred to in Section 2.4) for the General Electric LM 2500 [CBT]$_{IC}$ plant,

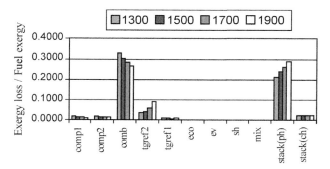

Fig. 5.10. Calculated exergy losses as fractions of fuel exergy for the General Electric LM 2500 [CBT] plant, for varying combustion temperatures (K) (after Ref. [13]).

for varying combustion temperatures. For the design T_{cot} of 1500 K the rational efficiency was calculated as 0.352 and the sum of all the fractional irreversibilities shown in the figure plus 0.352 thus gives unity. There are two major irreversibilities—that in combustion and the (physical) exergy loss in the stack gas due to its high temperature. (The 'chemical' exergy loss shown is that associated with the exergy theoretically available in the partial pressures of the exhaust, relative to atmosphere, as explained in Ref. [14].

The exergy losses in the HP turbine, which include losses in turbine cooling, are not negligible; those in the LP turbine are very small, since there is little or no cooling. Note, however, that it is the *total* turbine exergy losses that are shown here; reference should be made to the work of Young and Wilcock [15] for a detailed breakdown of such cooling exergy losses, into those associated with heat transfer, coolant throttling and mixing separately.

Fig. 5.11 shows the exergy losses as fractions of fuel exergy for the Westinghouse/Rolls-Royce WR21 recuperated [CICBTX]$_I$ plant. Now the stack (physical) exergy loss is much reduced by the action of the heat exchanger although the unit itself is not highly irreversible. At the design value of $T_{cot} = 1500$ K the rational efficiency is 0.371, which

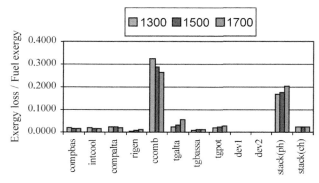

Fig. 5.11. Calculated exergy losses as fractions of fuel exergy for the Westinghouse/Rolls-Royce WR21 recuperated [CICBTX] plant, for varying combustion temperatures (K) (after Ref. [13]).

with all the irreversibilities shown sums to unity again. The combustion loss remains high at some 30%, and the HP turbine loss is not negligible.

5.9. Conclusions

In practice, the attainment of maximum thermal efficiency in a CBT gas turbine plant will depend on a complex mix of factors in addition to those for an uncooled plant, such as combustion temperature, pressure ratio and component efficiencies. The factors introduced by turbine cooling include the number of cooling steps, the quantities of cooling air required (crucially dependent on stagnation temperature at entry to each step, the permissible blade temperature and the temperature of the available cooling air), and the associated mixing losses. In addition, the properties of the working fluids (as real gases) also play an important part.

References

[1] Holland, M.J. and Thake, T.F. (1980), Rotor blade cooling in high pressure turbines, AIAA J. of Aircraft 17(6), 412–418.

[2] Horlock, J.H., Watson, D.T. and Jones, T.V. (2001), Limitations on gas turbine performance imposed by large turbine cooling flows, ASME J. Engng Gas Turbines Power 123, 4.

[3] Wilcock, R.C., Young, J.B. and Horlock, J.H. (2002), Gas properties as a limit to gas turbine performance, ASME paper GT-2002-30517.

[4] El-Masri, M.A. and Pourkey, F. (1986), Prediction of cooling flow requirements for advanced utility gas turbines, Part 1: Analysis and scaling of the effectiveness curve, ASME paper 86-WA/HT-43.

[5] El-Masri, M.A. (1986a), Prediction of cooling flow requirements for advanced utility gas turbines, Part 2: Influence of ceramic thermal barrier coatings, ASME paper 86-WA/HT-44.

[6] El-Masri, M.A. (1986b), On thermodynamics of gas turbine cycles: Part I second law analysis of combined cycles, ASME J. Engng Power Gas Turbines 107, 880–889.

[7] El-Masri, M.A. (1986c), On thermodynamics of gas turbine cycles: Part II Model for expansion in cooled turbines, ASME J. Engng Power Gas Turbines 108, 151–159.

[8] El-Masri, M.A. (1988), GASCAN—an interactive code for thermal analysis of gas turbine systems, ASME J. Engng Power Gas Turbines 110, 201–209.

[9] El-Masri, M.A. (1987a), Exergy analysis of combined cycles: Part 1 Air-cooled Brayton-cycle gas turbines, ASME J. Engng Power Gas Turbines 109, 228–235.

[10] El-Masri, M.A. (1987b), Exergy analysis of combined cycles: Part 2. Steam bottoming cycles, ASME J. Engng Power Gas Turbines 109, 237–243.

[11] Chiesa, P., Consonni, S., Lozza, G. and Macchi, E. (1993), Predicting the ultimate performance of advanced power cycles based on very high temperatures, ASME Paper 93-GT-223.

[12] MacArthur, C. D. (1999), Advanced aero-engine turbine technologies and their application to industrial gas turbines, ISABE Paper No. 99-7151, 14th International Symposium on Air-Breathing Engines, Florence, Italy, 1999.

[13] Facchini, B., Fiaschi, D. and Manfrida, G. (2000), Exergy analysis of combined cycles using latest generation gas turbines, ASME J. Engng Gas Turbines Power 122, 233–238.

[14] Horlock, J.H., Manfrida, G. and Young, J.B. (2000), Exergy analysis of modern fossil-fuel power plants, ASME J. Engng Gas Turbines Power 122, 1–17.

[15] Young, J.B. and Wilcock, R.C. (2002), Modelling the air-cooled gas turbine. Part 1—General thermodynamics, ASME J. Turbomachinery 124, 207–213.

Chapter 6

'WET' GAS TURBINE PLANTS

6.1. Introduction

As Frutschi and Plancherel [1] have explained, there are two basic gas turbine plants with water injection; they are illustrated in Fig. 6.1.

Fig. 6.1a shows diagrammatically the steam injection gas turbine (STIG) plant; *steam*, raised in a heat recovery steam generator (HRSG) downstream of the turbine, is injected into the combustion chamber or into the turbine nozzle guide vanes.

Fig. 6.1b shows diagrammatically the evaporative gas turbine (EGT) in which *water* is injected into the compressor outlet and is evaporated there; the mixture may then be further heated in the 'cold' side of a heat exchanger. It enters the combustion chamber and then passes through the turbine and the 'hot' side of the heat exchanger.

There are many variations on these two basic cycles which will be considered later. But first we discuss the basic thermodynamics of the STIG and EGT plants.

6.2. Simple analyses of STIG type plants

6.2.1. The basic STIG plant

Fig. 6.2 shows a simplified diagram of the basic STIG plant with steam injection S per unit air flow into the combustion chamber; the state points are numbered. Lloyd [2] presented a simple analysis for such a STIG plant based on 'heat input', work output and 'heat rejected' (as though it were a closed cycle air and water/steam plant, with external heat supplied instead of combustion and the exhaust steam and air restored to their entry conditions by heat rejection). His analysis is adapted here to deal with an open cycle plant with a fuel input f to the combustion chamber per unit air flow, at ambient temperature T_0, i.e. a fuel enthalpy flux of fh_{f0}. For the combustion chamber, we may write

$$h_{a2} + fh_{f0} + Sh_{s6} = (1+f)h_{g3} + Sh_{s3}, \tag{6.1}$$

where subscripts a, g and s refer to air, gas (products of combustion) and steam. The enthalpy of the steam quantity (h_s) is at the same temperature as the gas, and for convenience is carried separately through the analysis, i.e. the total enthalpy is $H = (1+f)h_g + Sh_s$. In reality, the steam and gas are fully mixed at all stations downstream of the combustion process.

Basic STIG

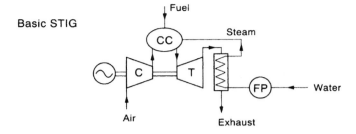

Evaporation plant

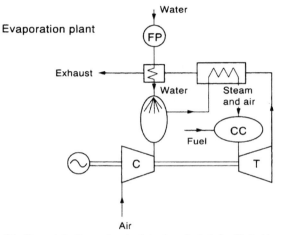

Fig. 6.1. Steam injection and water injection plants (after Frutschi and Plancherel [1]).

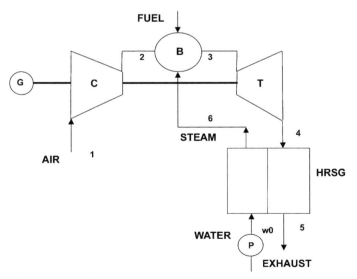

Fig. 6.2. Basic STIG plant (after Lloyd [2]). Princeton University Library.

In an experiment to determine the calorific value of the fuel at temperature T_0, and for the same fuel flow the steady flow energy equation would yield

$$h_{a0} + f h_{f0} = f[CV]_0 + (1 + f) h_{g0}. \tag{6.2}$$

Subtracting Eq. (6.2) from Eq. (6.1) yields

$$h_{a2} - h_{a0} + f[CV]_0 = (1 + f)(h_{g3} - h_{g0}) + S(h_{s3} - h_{s6}). \tag{6.3}$$

If the compressor entry temperature T_1 is the same as the ambient temperature T_0 then Eq. (6.3) may be rewritten as

$$(h_{a2} - h_{a1}) + f[CV]_0 = (1 + f)[(h_{g3} - h_{g4}) + (h_{g4} - h_{g5}) + (h_{g5} - h_{g0})] \\ + S[(h_{s3} - h_{s4}) + (h_{s4} - h_{s5}) + (h_{s5} - h_{s6})]. \tag{6.3a}$$

But across the HRSG the heat balance is

$$(1 + f)[(h_{g4} - h_{g5}) + S(h_{s4} - h_{s5})] = S(h_{s6} - h_{w0}), \tag{6.4}$$

in which the pumping work for the water is ignored, and the water enters at ambient temperature with enthalpy h_{w0}.

Combining this equation with Eq. (6.3a) yields the final energy equation for the whole plant as

$$f[CV]_0 = (W_T - W_C) + [(1 + f)(h_{g5} - h_{g0}) + S(h_{s5} - h_{w0})], \tag{6.5}$$

in which the terms in brackets correspond to the three terms in Lloyd's closed cycle analysis, Q_B, W, Q_A, respectively, and

$$Q_B = W + Q_A. \tag{6.6}$$

The overall efficiency of the plant is

$$\eta = (W_T - W_C)/f[CV]_0$$

$$= \{[1 + f][h_{g3} - h_{g4}] + S\{[h_{s3} - h_{s4}] - [h_{a2} - h_{a1}]\}/f[CV]_0, \tag{6.7}$$

so that

$$\eta = \left\{ 1 + \frac{(1 + f(h_{g5} - h_{g0}) + S(h_{s5} - h_{w0})}{(1 + f(h_{g3} - h_{g4}) + S(h_{s3} - h_{s4}) - (h_{a2} - h_{a1})} \right\}^{-1}, \tag{6.8}$$

by analogy with the form given by Lloyd,

$$\eta = W/Q_B = [1 + (Q_A/W)]^{-1}. \tag{6.9}$$

Lloyd argues that for a plant with fixed pressure ratio and top temperature, the turbine work output (and hence the net work output) is increased linearly with the steam quantity S that is injected, but the Q_B and Q_A terms increase more slowly. Thus, the efficiency similarly increases with S, but also more slowly.

Fig. 6.3, which gives illustrative plots of temperature against the fraction of heat transferred, shows how the HRSG performs, first at low S (Fig. 6.3a), and then with higher (optimum) S (Fig. 6.3b). Lloyd concludes that maximum efficiency is reached when

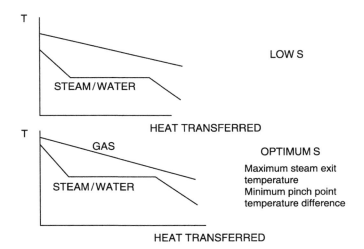

Fig. 6.3. HRSG performance of STIG plant at different steam/air ratios (after Lloyd [2]). Princeton University Library.

the superheated steam leaves at its upper temperature limit and when the pinch point temperature difference is at its minimum, as in diagram Fig. 6.3b.

He gave an example of an industrial gas turbine with a pressure ratio of 12 and a maximum temperature of 1100°C. For the basic CBT plant the specific work is approximately $W = W_T - W_C = 650 - 350 = 300$ kJ/kg (air) and $Q_B = 870$, so that $Q_A = 570$ and the efficiency is $300/870 = 0.345$. For a STIG plant with $S = 0.1$, the turbine work output increases by about 20% to 770 giving a net work output of 420, an increase of 40%. Q_B also increases somewhat, by about 23% to about 1070, and Q_A by about 14% to 650. The efficiency ($\eta = 420/1070$) therefore increases to nearly 40%, because the work output increases substantially more than the 'heat supplied'.

Fig. 6.4 then shows a more complete calculation of plant efficiency for varying S. The optimum condition of maximum efficiency is reached at $S = 0.208$. The picture changes for a gas turbine with a higher pressure ratio, for which the increase to maximum efficiency is less, as is the optimum value of S [2].

A useful rule of thumb is that the turbine work in a STIG plant is increased by a factor of about $(1 + 2S)$, since the specific heat of the steam is about double that of the specific heat of the 'dry' gas. This is in agreement with the example given above and with the earlier detailed calculations by Fraize and Kinney [3]. (Their work was based on the assumption that the mixture of air and steam in the turbine behaved as a semi-perfect gas, with specific heats being determined simply by mass averaging of the values for the two components.)

Finally, it may be noted that there is little or no point in adding steam directly to the turbine alone—say into the first nozzle guide vane row—because its enthalpy even at best would only be equal to the enthalpy of the steam leaving the turbine ($h_{s6} \leq h_{s4}$).

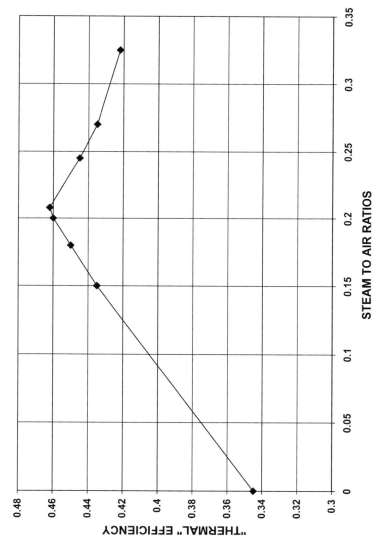

Fig. 6.4. Effect of steam air ratio (S) on STIG 'thermal' efficiency (after Lloyd [2]). Princeton University Library.

6.2.2. *The recuperative STIG plant*

Consider next a recuperative STIG plant (Fig. 6.5, again after Lloyd [2]). Heat is again recovered from the gas turbine exhaust: but firstly in a recuperator to heat the compressed air, to state 2A before combustion; and secondly in an HRSG, to raise steam S for injection into the combustion chamber.

Again we analyse the open cycle version of this plant, but with a fuel input f' (per unit air flow) at ambient temperature T_0, i.e. a fuel enthalpy flux of $f'h_{f0}$. For the combustion chamber we may now write

$$h_{a2A} + f'h_{f0} + Sh_{s6} = (1 + f')h_{g3} + Sh_{s3}, \tag{6.10}$$

and for the parallel calorific value experiment, at temperature $T_0 = T_1$,

$$h_{a0} + f'h_{f0} = f'[CV]_0 + (1 + f')h_{g0} \tag{6.11}$$

Subtracting Eq. (6.11) from Eq. (6.10) yields

$$h_{a2A} - h_{a0} + f'[CV]_0 = (1 + f')(h_{g3} - h_{g0}) + S(h_{s3} - h_{s6}), \tag{6.12}$$

There are now two heat balances: in the recuperator,

$$(1 + f')(h_{g4} - h_{g4A}) + S(h_{s4} - h_{s4A}) = (h_{a2A} - h_{a2}); \tag{6.13}$$

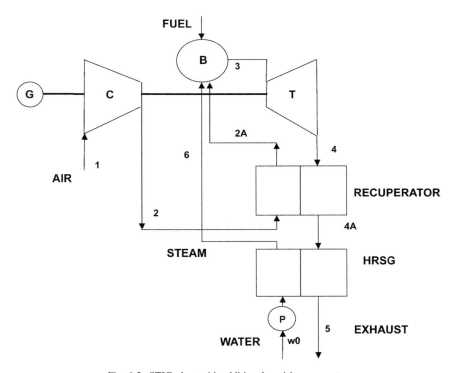

Fig. 6.5. STIG plant with additional gas/air recuperator.

and in the HRSG

$$(1 + f')(h_{g4A} - h_{g5}) + S(h_{s4A} - h_{s5}) = S(h_{s6} - h_{w0}). \tag{6.14}$$

But of course these two equations may be combined with Eq. (6.12) to give the steady flow energy equation for the whole plant as

$$f'[CV]_0 = (W_T - W_C) + (1 + f')(h_{g5} - h_{g0}) + S(h_{s5} - h_{w0}), \tag{6.15}$$

and the overall efficiency of the plant as

$$\eta = (W_T - W_C)/f'[CV]_0$$

$$= [(1 + f')(h_{g3} - h_{g4}) + S(h_{s3} - h_{s4}) - (h_{a2} - h_{a1})]/f'[CV]_0,$$

so that

$$\eta = \left\{1 + \frac{(1 + f')(h_{g5} - h_{g0}) + S(h_{s5} - h_{w0})}{(1 + f')(h_{g3} - h_{g4}) + S(h_{s3} - h_{s4}) - (h_{a2} - h_{a1})}\right\}^{-1}. \tag{6.16}$$

Eq. (6.16) is essentially the same as Eq. (6.8) for the basic STIG plant which, on reflection, is not surprising. If the states 1, 2, 3, 4 and 5 and the steam quantity S are all the same then expressions for the work output, the 'heat input' (or fuel energy supply) and the 'heat rejected' are all unchanged. The total amount of heat transferred from the exhaust is also unchanged, but two separate flows, of air and of water/steam, have been raised in enthalpy before entry to the combustion chamber, rather than one (water/steam) in the earlier analysis.

However in practice, for the same states 1–5 the steam raised S will be less; hence there is no advantage in operating a STIG plant in this variation of the basic CBTX recuperative gas turbine plant. Nonetheless, this form of analysis as developed by Lloyd will prove to be useful in the discussion of the chemical recuperation plant in Chapter 8.

6.3. Simple analyses of EGT type plants

6.3.1. A discussion of dry recuperative plants with ideal heat exchangers

Before considering the effects of water injection in an EGT type plant, it is worthwhile to refer to the earlier studies on the performance of some dry recuperative cycles. Fig. 6.6 shows the T, s diagram of a $[CBT]_I[X]_R$ cycle, with a heat exchanger effectiveness of unity. It is implied that the surface area for heat transfer is very large, so that the outlet temperature on the cold side is the same as the inlet temperature on the hot side. However, due to the higher specific heat of the hot gas, its outlet temperature is higher than the inlet temperature of the cold air.

In their original air standard cycle analysis, using constant specific heats, Hawthorne and Davis [4] considered the dry $[CBT]_I X_R$ cycle. They assumed a 'perfect' heat exchanger, with the specific heats of gas and air constant and identical, so that T_Y becomes equal to T_2 in Fig. 6.6. From their examination of the enthalpy–entropy diagram of this

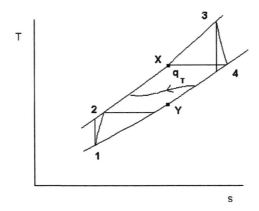

Fig. 6.6. Temperature–entropy diagram for dry $[CHT]_I X_R$ plant.

cycle they then concluded the turbine work (W_T) to be equal to the heat supplied (Q_B) so the efficiency becomes

$$\eta = W/Q_B = (W_T - W_C)/W_T = 1 - (W_C/W_T). \tag{6.17}$$

Note also that the heat rejected is equal to the compressor work in this case. For this air standard cycle with constant specific heats, Eq. (6.17) reduces simply to

$$\eta = 1 - (x/\alpha), \tag{6.18a}$$

where $\alpha = \eta_C \eta_T (T_3/T_1)$, $x = r^{(\gamma-1)/\gamma}$ and η_C and η_T are isentropic efficiencies.

Consider next a similar recuperative cycle, but one in which the compression process approximates to isothermal rather than isentropic, with the work input equal to the heat rejected (this may be achieved in a series of small compressions of *polytropic* efficiency η_p, followed by a series of constant pressure heat rejections). It may then be shown that the thermal efficiency of this cycle is given by

$$\eta = 1 - \{\ln \phi/[\theta(1 - \phi^{-1/2})]\}, \tag{6.18b}$$

where $\theta = T_3/T_1$, $\phi = r^\sigma$ and $\sigma = (\gamma - 1)/\gamma\eta_p$. This cycle is more efficient than the $[CBT]_I X_R$ cycle, and this will be important when we consider its evaporative version later (the TOPHAT or van Liere cycle).

For the $(CICBT)_I X_R$, $(CBCBT)_I X_R$ and $(CICBTBT)_I X_R$ cycles, with equal pressure ratios across the 'split' compressors and turbines, it may be shown that the corresponding expressions for efficiency are

$$\eta = 1 - 2x/\alpha(x^{1/2} + 1), \tag{6.18c}$$

$$\eta = 1 - (x^{1/2} + x)/2\alpha, \tag{6.18d}$$

$$\eta = 1 - x^{1/2}/\alpha, \tag{6.18e}$$

respectively, indicating that the efficiency increases with α in each of these cycles.

The thermal efficiencies (η) of these five cycles, all with perfect recuperation, are plotted in Fig. 6.7 against the isentropic temperature ratio x, for $\eta_T \eta_C = 0.8$ and $T_3/T_1 = 5$

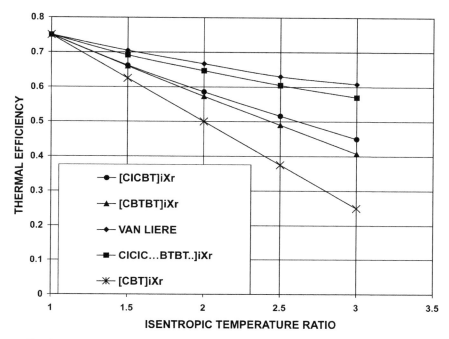

Fig. 6.7. Air standard thermal efficiencies of various dry plants with reversible recuperators.

($\alpha = 4$), but with $\eta_p = 0.9$ for the van Liere cycle. The thermal efficiency η of each cycle is highest for $x \to 1.0$, for which it is equal to $[1 - (1/\alpha)]$. For the $[CBT]_I X_R$ cycle η decreases rapidly with increasing pressure ratio; the efficiency of the other cycles drops less rapidly. Reheating and intercooling raises the efficiency but the thermal efficiency of the van Liere cycle is highest. It drops slowly with x, but its efficiency is almost matched by the cycle with both reheating and intercooling.

In practice, however, the heat exchanger effectiveness will not be unity for these dry cycles, but the above analysis does suggest that for practical plants:

(i) the optimum pressure ratio for a $[CBTX]_I$ plant will be low (as was illustrated in Fig. 3.15, for a realistic heat exchanger effectiveness of 0.75);

(ii) the introduction of intercooling and reheating will increase the efficiency in the recuperative cycles and also raise the optimum pressure ratio.

6.3.2. The simple EGT plant with water injection

The discussion of the last section is then useful in considering the evaporative cycles. We shall see that the effect of water injection downstream of the compressor (and possibly in the cold side of the heat exchanger) may lead towards the $[CBT]_I X_R$ type of plant, with increased cold side effective specific heat and hence increased heat exchanger effectiveness. Water injection in the compressor may lead to a plant with isothermal compression.

Firstly, Fig. 6.8a shows the T, s (air property) diagram for an EGT cycle, a 'wet' version of the CBTX cycle with water injected to cool the compressor discharge air. Frutschi and Plancherel argued that the virtue of such evaporative cooling before the heat exchanger is to drop the hot gas temperature at the exchanger exit. A closed cycle version of this EGT cycle, in which the water injected was condensed after exit from the heat exchanger and then recirculated to complete the cycle, was initially considered by Horlock [5]. This analysis showed that the temperature of the gas at exit from the heat exchanger was indeed reduced in the wet cycle; the total heat rejected (Q_A) was unchanged from that in the dry cycle, because of the condensation of the steam which was necessary to close the wet cycle. Some of the heat rejected in the dry cycle is simply moved from the gas flow downstream of the hot side of the heat exchanger to the additional condenser required in the wet cycle.

However, the turbine work has been increased because of the extra water vapour flow through the turbine, while the compressor work is unchanged. Thus Eq. (6.17), which is still valid, with turbine work equal to the heat supplied, shows that the thermal efficiency increases compared with the dry cycle. It is important to realise that this efficiency is increased not because of a reduction in the heat rejected (Q_A) but because of the increase in W_T. The heat rejected is still equal to the compressor work.

If, as suggested in Section 6.2.1, the turbine work is increased by a factor $(1 + 2S)$, where S is the water vapour flow, then the dry and wet efficiencies may be written as

$$\eta_{DRY} = 1 - (W_C/W_{T\ DRY}), \tag{6.19a}$$

and

$$\eta_{WET} = 1 - [W_C/(1 + 2S)(W_{T\ DRY}], \tag{6.19b}$$

so that

$$(\eta_{WET} - \eta_{DRY})/(1 - \eta_{DRY}) \approx 2S/(1 + 2S) \tag{6.19c}$$

The same expression applies for some of the other variations of the EGT cycle considered below (e.g. the recuperative water injection (RWI) plant with intercooling).

Horlock then considered a cycle proposed by El-Masri [6] in which the water evaporation takes place not in an aftercooler but in the cold side of the heat exchanger; again a cycle in closed form was considered, with injected water finally condensed and recirculated. The continuing evaporation increases the effective specific heat of the cold side fluid and can increase the effectiveness of the heat exchanger towards unity [7]. The Hawthorne and Davis analysis for the dry $[CBT]_I X_R$ cycle (and also for the other more complex dry cycles) then becomes relevant. For a 'perfect' heat exchanger in the closed EGT cycle, in which the continued evaporation on the cold side can lead to the hot and cold side specific heats becoming the same, the heat rejected is now equal to the compressor work. The temperature–entropy diagram for the 'carrying' gas is now shown in Fig. 6.8b. The expression for the air standard efficiency of the closed dry CBTX cycle (Eq. (6.17)) is also valid for this EGT cycle, with $Q_A = W_C$, the value in a dry cycle. But the turbine work W_T ($= Q_B$) is increased because of the extra steam passing through the turbine, with its associated enthalpy drop. Again this is the essence of the EGT cycle where

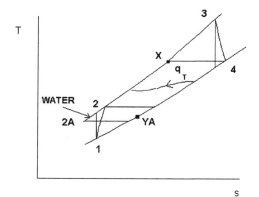

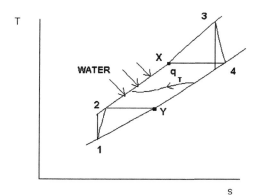

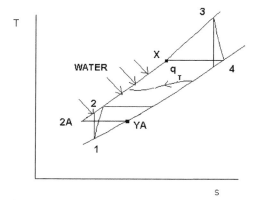

Fig. 6.8. (a) Temperature–entropy diagram for water injection into aftercooler of $[CHT]_IX_R$ plant (after Ref. [5]).
(b) Temperature–entropy diagram for water injection into cold side of heat exchanger of $[CHT]_IX_R$ plant.
(c) Temperature–entropy diagram for water injection into aftercooler and cold side of heat exchanger of $[CHT]_IX_R$ plant (after Ref. [5]).

an increase in turbine work (and heat supplied) with a constant compressor work (and heat rejected), leads to an increase in efficiency.

A further variation of the El-Masri EGT cycle is one in which the evaporation takes place both in an aftercooler and within the cold side of the heat exchanger (Fig. 6.8c). Eq. (6.17) is still valid, but the efficiency is increased because more water can be injected and the turbine work increased further.

It was shown in Ref. [5] that the arguments given above for the closed EGT cycles also hold good for open EGT cycles, but this analysis is not repeated here. Some simple parametric calculations were given to illustrate the increased thermal efficiency of practical open EGT cycles, corresponding to Fig. 6.8a–c. It was assumed that water injection was

(a) in the aftercooler (sufficient to saturate the compressor discharge gas),
(b) within the heat exchanger (cold side, to raise the effective specific heat), and
(c) in both aftercooler and heat exchanger (cold side).

Evaporative mixing at low velocity was assumed in the aftercooler, the pressure remaining constant. Allowance was made for the real gas effects (increased specific heat of the products of combustion at high temperature), of turbine cooling and intercooling. A method of calculating the turbine work similar to that developed by Cerri and Arsuffi [8] for the STIG cycle was used. It was assumed that evaporative cooling was carried out by a small quantity of water so that the temperatures of the working gas carrying the steam remain unchanged (except after injection for intercooling and at exit from the hot side of the heat exchanger). The additional steam in the turbine was assumed to be superheated (at low partial pressure) and its drop in enthalpy was obtained from steam tables knowing the original 'dry' gas temperature drop.

Plots of efficiency against pressure ratio for the full injection EGT plant, for a maximum to minimum temperature ≈ 5, are shown in Fig. 6.9, compared with lower values of efficiency in the dry CBTX plant. There are several points to be noted: first that an increase in efficiency is worthwhile, up to 10%; secondly that the total water injection is up to over 10% of the air mass flow; and thirdly that the optimum pressure ratio increases to about 8, from about 5 for that of the dry cycle.

Similar calculations (Fig. 6.10) were made for intercooled cycles, without and with water injection, i.e. comparing the efficiency of the dry CICBTX cycle with an elementary recuperated water injection plant, now a simple version of the so-called RWI plant (see Section 6.4.2.1). Again there is an increase in thermal efficiency with water injection, but it is not as great as for the simple EGT plant compared with the dry CBTX plant; the optimum pressure ratio, about 8 for the dry intercooled plant, appears to change little with water injection.

These smaller effects are related to the smaller amount of water that can be injected for the intercooled cycle. Applying Eq. (6.19c) to the near optimum condition of Fig. 6.9 ($\eta_{DRY} = 0.5$, with $S = 0.1$) yields ($\eta_{WET} - \eta_{DRY}$) ≈ 0.08. Applying the same equation to the near optimum condition of Fig. 6.10 ($\eta_{DRY} = 0.53$, with $S = 0.04$) yields ($\eta_{WET} - \eta_{DRY}$) ≈ 0.035. Both these approximate estimates are very close to the detailed calculations of the increases in thermal efficiency shown in the two figures.

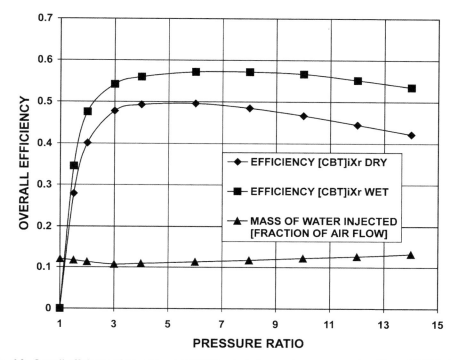

Fig. 6.9. Overall efficiency of dry and wet [CBT]$_I$X$_R$ plants for varying pressure ratios ($T_{cot} = 1200°C$) (after Ref. [5]).

All these calculations showed modest increases in specific work, consistent with the relatively small amounts of water injection.

6.4. Recent developments

Several modifications of the two basic steam and water injection plants (STIG and EGT) have been proposed in recent years. Rather than analysing all these developments in detail here, they are first briefly described; subsequently the 'thermodynamic intentions' of these modified cycles are discussed, together with some of the parametric studies which have been made by other authors.

6.4.1. Developments of the STIG cycle

6.4.1.1. The ISTIG cycle

A development of the STIG cycle is the intercooled steam injection cycle (ISTIG) shown in Fig. 6.11. It involves raising steam through the waste heat of a basic CICBT plant and injecting it into the combustion chamber; the intercooling may be by surface intercooling or evaporative intercooling. The essential feature of the ISTIG proposal is to obtain increased turbine work in a cycle which already has a large specific work because of the intercooling.

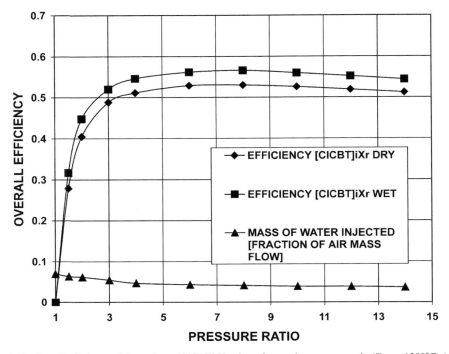

Fig. 6.10. Overall efficiency of dry and wet $[CICBT]_IX_R$ plants for varying pressure ratio ($T_{cot} = 1200°C$) (after Ref. [5]).

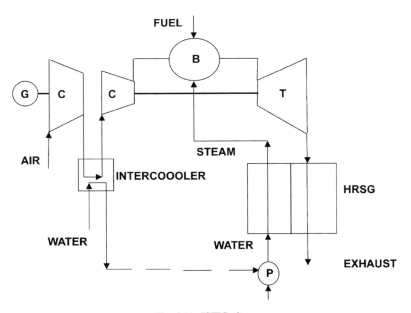

Fig. 6.11. ISTIG plant.

Combined STIG

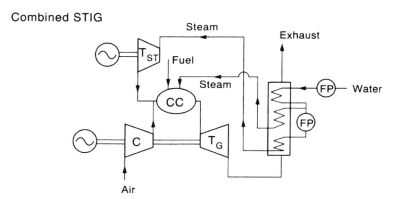

Fig. 6.12. Combined STIG plant (after Frutschi and Plancherel [1]).

6.4.1.2. The combined STIG cycle

The combined STIG cycle (Fig. 6.12) was described by Frutschi and Plancherel [1]. Steam is raised at two pressure levels in the waste heat boiler. Superheated steam at the higher pressure level expands through a steam turbine before injection into the compressor discharge air stream. Low pressure steam is injected (STIG fashion) into the combustion chamber. Attainable efficiency for this plant may in theory reach about 50%. In a variation of this combined cycle (the Foster–Pegg plant), the steam turbine drives a second high pressure compressor.

6.4.1.3. The FAST cycle

Another modification of the combined STIG cycle is the so-called advanced steam topping (FAST) cycle. Now the double steam injection process (before and after combustion) of the combined STIG cycle of Fig. 6.12 is replaced by a single steam injection into the combustion chamber, after expansion in the steam turbine and reheating in the HRSG (Fig. 6.13). In one version the steam turbine and the gas turbine are on the same shaft, jointly driving the electrical generator. To call this cycle a steam topping cycle is somewhat misleading, since it is essentially a doubly open combined cycle in that heat rejection from the (upper) gas turbine is rejected to a (lower) main steam turbine cycle. This lower cycle now includes reheating, steam leaving the steam turbine being reheated before a second expansion in the gas turbine. But, of course, the steam is exhausted with the gas and is not finally condensed, and there is no recirculation of water.

6.4.2. Developments of the EGT cycle

There have been a larger number of proposals for recuperated cycles with water injection and evaporation, but all these can be interpreted as modifications of the EGT plant, which is essentially a 'wet' CBTX cycle, as explained above.

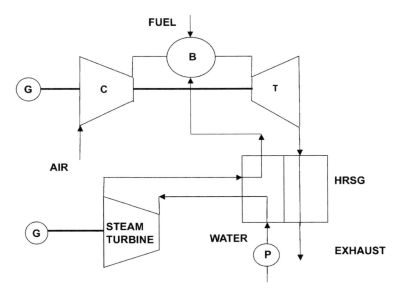

Fig. 6.13. Advanced steam topping (FAST) plant.

6.4.2.1. The RWI cycle

Frutschi and Plancherel [1] not only described the basic EGT cycle, but also a modified version with an intercooler added. Macchi et al. [9] called this intercooled EGT the RWI plant and the simplest version is shown in the top part of Fig. 6.14. Macchi et al. also considered more complex versions (some with evaporative intercooling and aftercooling), the performance of which are discussed in Section 6.6.

6.4.2.2. The HAT cycle

A further major innovation is the humidified air turbine (HAT) cycle, which involves introduction of a humidifier before the combustion chamber, rather than the mixer originally proposed by Frutschi and Plancherel. The resulting HAT cycle is shown diagrammatically, as a modification of the simply intercooled RWI cycle, in the lower part of Fig. 6.14. There is now a smaller exergy loss in the evaporation process, both from increasing the water temperature at entry to the humidifier (by using cooling water passing through the intercoolers between LP and HP compressors and an aftercooler), and from reduction of the temperature difference between the water and air within the humidifier itself.

6.4.2.3. The REVAP cycle

De Ruyck et al. [10] proposed another variation of the EGT cycle, in an attempt to reduce the exergy losses involved in water injection (the REVAP cycle). Rather than introducing the complication of a saturator, De Ruyck proposed several stages of water heating (in an economiser, an intercooler and an aftercooler). The efficiency claimed for this cycle is only a little less than the HAT cycle.

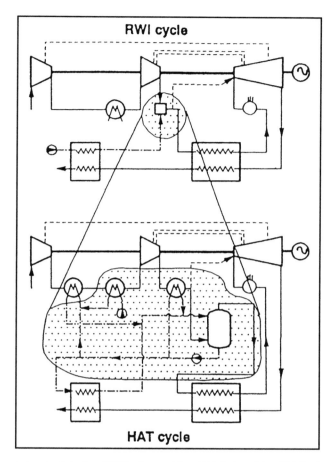

Fig. 6.14. Recuperated water injection (RWI) plant and humidified air turbine (HAT) plant compared (after Macchi et al. [9]).

6.4.2.4. *The CHAT cycle*

A modification of the HAT cycle has been proposed by Nakhamkin [11], which is known as the cascaded humid air turbine (CHAT). The higher pressure ratios required in humidified cycles led Nakhamkin to propose reheating between the HP and LP turbines. Splitting the expansion in this way is paralleled by splitting the compression, and enables the HP shaft to be non-generating, as indicated in Fig. 6.15. This implies that the capital cost of the plant can be reduced, but the cycle is still complex.

6.4.2.5. *The TOPHAT cycle*

Another water injection cycle proposed is the TOPHAT cycle [12] (see Fig. 6.16). As for the HAT cycle, the purpose is to introduce water into the cycle with low exergy loss and this is achieved by injecting water continuously in the compressor in an attempt to

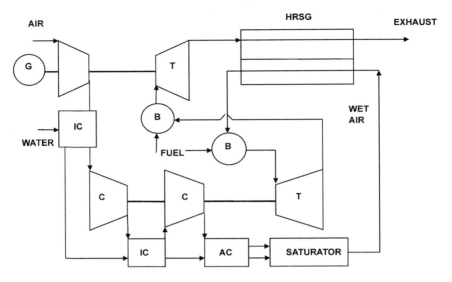

Fig. 6.15. Cascaded humid air turbine (CHAT) plant.

move the compression towards isothermal rather than adiabatic, with the consequence of reduced work input. Now the claim is for an efficiency higher than that of the HAT cycle, and this may be expected from the analysis of the dry 'van Liere' cycle given in Section 6.3.1.

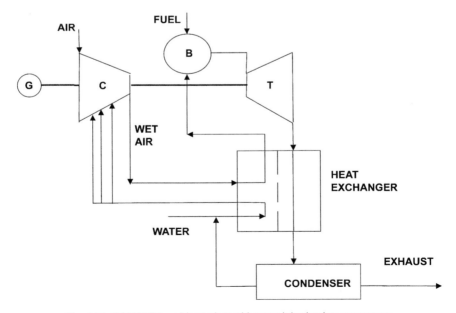

Fig. 6.16. TOPHAT (van Liere) plant with water injection into compressor.

6.4.3. Simpler direct water injection cycles

In the search for higher plant thermal efficiency, the simplicity of the two basic STIG and EGT cycles, as described by Frutschi and Plancherel, has to some extent been lost in the substantial modifications described above. But there have been other less complex proposals for water injection into the simple unrecuperated open cycle gas turbine; one simply involves water injection at entry to the compressor, and is usually known as inlet fog boosting (IFB); the other involves the 'front part' of an RWI cycle, i.e. water injection in an evaporative intercooler, usually in a high pressure ratio aero-derivative gas turbine plant.

For the IFB plant the main advantage lies in the reduction of the inlet temperature, mainly by saturating the air with a very fine spray of water droplets [13]. This, in itself, results in an increased power output, but it is evident that the water may continue to evaporate within the compressor, resulting in a lowering of the compressor delivery temperature. A remarkable result observed by Utamura is an increase of some 8% in power output for only a small water mass flow (about 1% of air mass flow). However, the compressor performance may be adversely affected as the stages become mismatched [14], even for the small water quantities injected.

In the second development, the emphasis is on taking advantage of the increased specific work associated with evaporative intercooling and of the increased mass flow and work output of the turbine. Any gain on the dry efficiency is likely to be marginal, depending on the split in pressure ratio.

6.5. A discussion of the basic thermodynamics of these developments

All these cycles involve attempts to improve on the various 'dry' gas turbine cycles discussed earlier in Section 6.3.

The basic STIG cycle improves on the dry CBT cycle through an element of recuperation and by increasing the turbine work [2]. The ISTIG cycle provides a similar improvement of the dry CICBTX cycle with the extra flow through the turbine. The combined STIG and FAST cycles involve introducing a steam turbine giving extra work and move the simple STIG cycle into the realms of the combined cycle plant (see Chapter 7).

To further understand the 'thermodynamic philosophy' of the improvements on the EGT cycle we recall the cycle calculations of Chapter 3 for ordinary dry gas turbine cycles—including the simple cycle, the recuperated cycle and the intercooled and reheated cycles.

Fig. 3.16 showed carpet plots of efficiency and specific work for several dry cycles, including the recuperative [CBTX] cycle, the intercooled [CICBTX] cycle, the reheated [CBTBTX] cycle and the intercooled reheated [CICBTBTX] cycle. These are replotted in Fig. 6.17. The ratio of maximum to minimum temperature is 5:1 (i.e. $T_{max} \approx 1500$ K); the polytropic efficiencies are 0.90 (compressor), 0.88 (turbine); the recuperator effectiveness is 0.75. The fuel assumed was methane and real gas effects were included, but no allowance was made for turbine cooling.

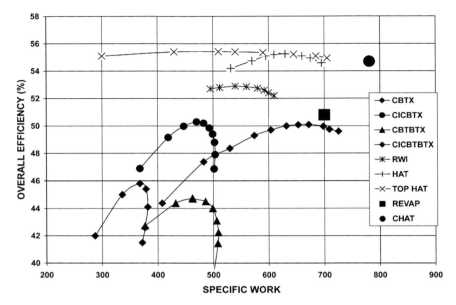

Fig. 6.17. Overall efficiency and specific work of dry and wet cycles compared.

To this figure, some of the calculations carried out by various authors for wet cycles have been added: RWI and HAT [9]; REVAP [10]; CHAT [11]; TOPHAT [12].

The assumptions made by the various authors (viz. polytropic efficiencies, combustion pressure loss and temperature ratio, etc.) are all roughly similar to those used in the calculations of uncooled dry cycles. Some modest amounts of turbine cooling were allowed in certain cases [9] but the effect of these on the efficiency should not be large at $T_{max} \approx 1250°C$ (see later for discussion of more detailed parametric calculations by some of these authors).

The RWI and HAT cycles may then be seen as 'wet' developments of the intercooled regenerative dry cycle. These evaporative cycles show an increase in efficiency on that of the dry CICBTX cycle—largely because of the increased turbine work (still approximately the same as the 'heat supplied') which is not at the expense of increased compressor work. The HAT cycle then offers an appreciable reduction in the exergy loss in the evaporative process compared with RWI, thus providing an added advantage in terms of the thermal efficiency. REVAP also provides a similar advantage on efficiency. The TOPHAT cycle has the advantage of increased turbine work together with reduced compressor work.

The CHAT cycle may be seen as a low loss evaporative development of the dry intercooled, reheated regenerative cycle [CICBTBTX]. It offers some thermodynamic advantage—increase in turbine work (and 'heat supplied') with little or no change in the compressor work, leading to an increased thermal efficiency and specific work output.

In summary, all these 'wet' cycles may be expected to deliver higher thermal efficiencies than their original dry equivalents, at higher optimum pressure ratios. The specific work quantities will also increase, depending on the amount of water injected.

6.6. Some detailed parametric studies of wet cycles

The general thermodynamic conclusions given above are confirmed by more detailed parametric studies which have been made by several authors of various wet cycles.

Macchi et al. [9] made an extensive study of water injection cycles in their two classic papers and their results are worth a detailed study. Some of their calculations (for ISTIG, RWI and HAT) are reproduced in Figs. 6.18–6.20, all for surface intercooling (parallel calculations for evaporative intercooling are given in the original papers).

For the ISTIG cycle, Fig. 6.18 shows thermal efficiency plotted against specific work for varying overall pressure ratios and two maximum temperatures of 1250 and 1500°C. Peak efficiency is obtained at high pressure ratios (about 36 and 45, respectively), before the specific work begins to drop sharply. Note that the pressure ratios of the LP and HP compressors were optimised within these calculations.

Macchi et al. provided a similar comprehensive study of the more complex RWI cycles as illustrated in Fig. 6.19, which shows similar carpet plots of thermal efficiency against specific work for maximum temperatures of 1250 and 1500°C, for surface intercoolers. The division of pressure ratio between LP and HP compressors is again optimised within these calculations, leading to an LP pressure ratio less than that in the HP. For the RWI cycle at 1250°C the optimisation appears to lead to a higher optimum overall pressure ratio (about 20) than that obtained by Horlock [5], who assumed LP and HP pressure ratios to be same in his study of the simplest RWI (EGT) cycle. His estimate of optimum pressure ratio

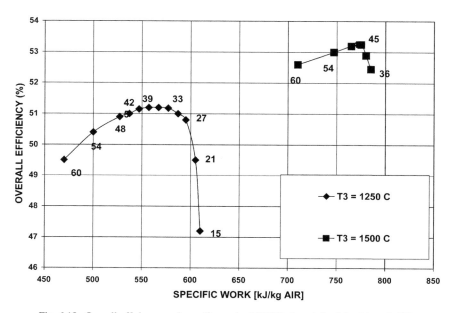

Fig. 6.18. Overall efficiency and specific work of ISTIG plant (after Macchi et al. [9]).

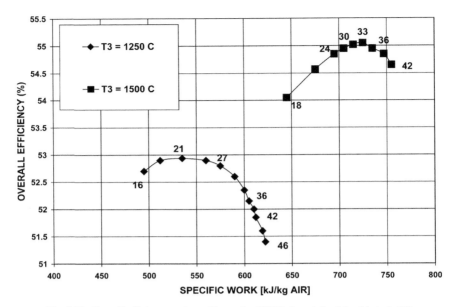

Fig. 6.19. Overall efficiency and specific work of RWI plant (after Macchi et al. [9]).

was in the region of 10, but the efficiency plot against pressure ratio was very flat, and of course the calculation method much simplified.

Macchi et al. presented similar calculations for the HAT cycle based on comparable assumptions (Fig. 6.20). As to be expected, they obtain efficiencies about 2% higher

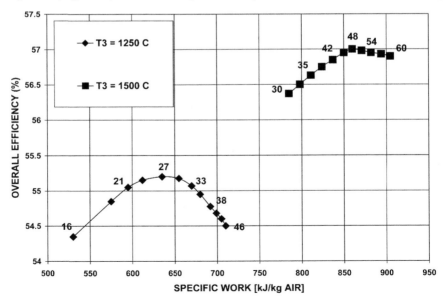

Fig. 6.20. Overall efficiency and specific work of HAT plant (after Macchi et al. [9]).

than the RWI calculations, peaking at even higher pressure ratios (27 at 1250°C, 50 at 1500°C).

Macchi et al. did not undertake parametric studies of the CHAT cycle and there appears to be no comparably thorough examination of this cycle in the literature; but Nakhamkin describes a prototype plant giving a thermal efficiency of some 55% at a very high pressure ratio, i.e. about 70, compared with the dry CICBTBTX cycle optimum of about 40 shown in Fig. 6.17.

van Liere's calculations for the TOPHAT cycle, also shown in Fig. 6.17, show a remarkably flat variation in efficiency for a wide variation in specific work.

6.7. Conclusions

The main conclusions from the work on water injection describes in this chapter are as follows:

(i)   the well established STIG cycle shows substantial improvement on the dry CBT cycle, mainly in specific work but also in thermal efficiency;

(ii) the simple EGT plant (a 'wet' CBTX cycle) cycle gives an increase in the thermal efficiency; the optimum pressure ratio is still quite low, but a little above that of the dry CBTX cycle;

(iii) the intercooled RWI, HAT, REVAP and TOPHAT cycles give increases of efficiency and specific work on the dry CICBTX cycle, at the expense of the added complexity, optimum conditions occurring at higher pressure ratios;

(iv) the CHAT cycle, interpreted as an evaporative modification of the 'ultimate' dry CICBTBTX plant, appears to yield high efficiency at an even higher pressure ratio.

References

[1] Frutschi, H.U. and Plancherel, A.A. (1988), Comparison of combined cycles with steam injection and evaporation cycles, Proc. ASME COGEN-TURBO II, pp.137–145.

[2] Lloyd, A. (1991), Thermodynamics of chemically recuperated gas turbines. CEES Report 256, Centre For Energy and Environmental Studies, University Archives, Department of Rare Books and Special Collections, Princeton University Library.

[3] Fraize, W.E. and Kinney, C. (1979), Effects of steam injection on the performance of gas turbines and combined cycles, ASME J. Engng Power Gas Turbines 101, 217–227.

[4] Hawthorne, W.R. and Davis, G.de V. (1956), Calculating gas turbine performance, Engineering 181, 361–367.

[5] Horlock, J.H. (1998), The evaporative gas turbine, ASME J. Engng Gas Turbines Power 120, 336–343.

[6] El-Masri, M.A. (1988), A modified high efficiency recuperated gas turbine cycle, J. Engng Gas Turbines Power 110, 233–242.

[7] Horlock, J.H. (1998), Heat exchanger performance with water injection (with relevance to evaporative gas turbine (EGT) cycles), Energy Conver Mgmt 39(16–18), 1621–1630.

[8] Cerri, G. and Arsuffi, G. (1986), Calculation procedures for steam injected gas turbine cycle with autonomous distilled water production, ASME Paper 86-GT-297.

[9] Macchi, E., Consonni, S., Lozza, G. and Chiesa, P. (1995), An assessment of the thermodynamic performance of mixed gas–steam cycles, Parts A and B, ASME J. Engng Gas Turbines Power 117, 489–508.

[10] De Ruyck, J., Bram, S. and Allard, G. (1997), REVAP cycle: A new evaporative cycle without saturation tower, ASME J. Engng Gas Turbines Power 119, 893–897.

[11] Nakhamkin, M., Swensen, E.C., Wilson, J.M., Gaul, G. and Polsky, M. (1996), The cascaded humidified advanced turbine (CHAT), ASME J. Engng Gas Turbines Power 118, 565–571.

[12] van Liere, J. (2001), The TOPHAT turbine cycle. Gas turbine technology, Modern Power Systems April, 35–37.

[13] Utamura, M., Takaaki, K., Murata, H. and Nobuyuki, H. (1999), Effects of intensive evaporative cooling on performance characteristics of land-based gas turbine, PWR-Vol. 34, Joint Power Generation Conference, pp. 321–328.

[14] Horlock, J.H. (2001), Compressor performance with water injection, ASME Paper 2001-GT-343.

Chapter 7

THE COMBINED CYCLE GAS TURBINE (CCGT)

7.1. Introduction

The modification to single cycles described earlier may not achieve a high enough overall efficiency. The plant designer therefore explores the possibility of using a combined plant, which is essentially one plant thermodynamically on top of the other, the lower plant receiving some or all of the heat rejected from the upper plant. If a higher mean temperature of heat supply and/or a lower temperature of heat rejection can be achieved in this way then a higher overall plant efficiency can also be achieved, as long as substantial irreversibilities are not introduced.

In this chapter, a short review of the thermodynamics of CCGTs is given. However, the author recommends readers to refer to two books which deal with combined plants in greater detail [1,2].

7.2. An ideal combination of cyclic plants

Consider a combined power plant made up of two cyclic plants (H, L) in series (Fig. 7.1). In this ideal plant, heat that is rejected from the higher (topping) plant, of thermal efficiency η_H, is used to supply the lower (bottoming) plant, of thermal efficiency η_L, with no intermediate heat loss and supplementary heating.

The work output from the lower cycle is

$$W_L = \eta_L Q_{HL}, \tag{7.1}$$

but

$$Q_{HL} = Q_B(1 - \eta_H), \tag{7.2}$$

where Q_B is the heat supplied to the upper plant, which delivers work

$$W_H = \eta_H Q_B. \tag{7.3}$$

Thus, the total work output is

$$W = W_H + W_L = \eta_H Q_B + \eta_L(1 - \eta_H)Q_B = Q_B(\eta_H + \eta_L - \eta_H \eta_L). \tag{7.4}$$

The thermal efficiency of the combined plant is therefore

$$\eta_{CP} = \frac{W}{Q_B} = \eta_H + \eta_L - \eta_H \eta_L. \tag{7.5}$$

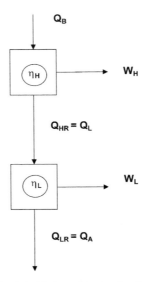

Fig. 7.1. Ideal combined cycle plant.

The thermal efficiency of the combined plant is greater than that of the upper cycle alone, by an amount $\eta_L(1 - \eta_H)$.

7.3. A combined plant with heat loss between two cyclic plants in series

Consider next two cyclic plants operating in series, but with unused heat Q_{UN} (or heat 'loss') between the two plants, so that $Q_{HR} = Q_L + Q_{UN}$, as shown in Fig. 7.2.

The overall thermal efficiency of the combined plant is by definition

$$\eta_{CP} = \frac{W_H + W_L}{Q_B},$$

and the efficiencies of the higher and lower plant, respectively, are

$$\eta_H = \frac{W_H}{Q_B}, \qquad \eta_L = \frac{W_L}{Q_L}.$$

However, the heat supplied to the lower cycle is now

$$Q_L = Q_{HR} - Q_{UN} = Q_B(1 - \eta_H) - Q_{UN},$$

so that

$$\eta_{CP} = \frac{\eta_H Q_B + \eta_L [Q_B(1 - \eta_H) - Q_{UN}]}{Q_B} = \eta_H + \eta_L - \eta_H \eta_L - v_{UN} \eta_L, \qquad (7.6)$$

where $v_{UN} = Q_{UN}/Q_B$. Thus there is a loss in efficiency of $v_{UN}\eta_L$, in comparison with the 'ideal' cycle with no heat loss between plants H and L.

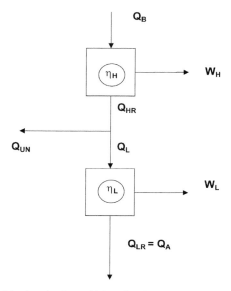

Fig. 7.2. Combined cycle plant with heat loss between higher and lower plants.

Eq. (7.6) can be written in another form, defining

$$\eta_B = \frac{Q_L}{Q_{HR}} = 1 - \frac{Q_{UN}}{Q_{HR}} = 1 - \left[\frac{\nu_{UN}}{(1 - \eta_H)} \right] \tag{7.7}$$

as the fraction of the total heat rejected by the higher cycle which is supplied to the lower cycle, a form of 'boiler efficiency' for the heat transfer process. The combined plant efficiency may be written as

$$\eta_{CP} = \eta_H + \frac{\eta_L Q_L}{Q_B} = \eta_H + \eta_L \eta_B \frac{Q_{HR}}{Q_B} = \eta_H + \eta_B \eta_L - \eta_B \eta_H \eta_L, \tag{7.8}$$

or

$$\eta_{CP} = \eta_H + (\eta_O)_L - \eta_H(\eta_O)_L, \tag{7.9}$$

where $(\eta_O)_L$ is the overall efficiency for the lower cycle, equal to the product of thermal efficiency and 'boiler efficiency', $\eta_L \eta_B$.

7.4. The combined cycle gas turbine plant (CCGT)

The most developed and commonly used combined power plant involves a combination of open circuit gas turbine and a closed cycle (steam turbine), the so-called CCGT. Many different combinations of gas turbine and steam turbine plant have been proposed. Seippel and Bereuter [3] provided a wide-ranging review of possible proposed plants, but essentially there are two main types of CCGT.

In the first type, heating of the steam turbine cycle is by the gas turbine exhaust with or without additional firing (there is normally sufficient excess air in the turbine exhaust for additional fuel to be burnt, without an additional air supply). In the second, the main combustion chamber is pressurised and joint 'heating' of gas turbine and steam turbine plants is involved.

Most major developments have been of the first (*exhaust heated*) system, with and without additional firing of the exhaust. The firing is usually 'supplementary'—burning additional fuel in the heat recovery steam generator (HRSG) up to a maximum temperature of about 750°C. However, full firing of exhaust boilers is used in the repowering of existing steam plants.

7.4.1. The exhaust heated (unfired) CCGT

Exhaust gases from the gas turbine are used to raise steam in the lower cycle without the burning of additional fuel (Fig. 7.3); the temperatures of the gas and water/steam flows are as indicated. A limitation on this application lies in the heat recovery system steam generator; choice of the evaporation pressure (p_c) is related to the temperature difference ($T_6 - T_c$) at the 'pinch point' as shown in the figure, and a compromise has to be reached between that pressure and the stack temperature of the gases leaving the exchanger, T_S (and the consequent 'heat loss').[1]

We first consider how the simple analysis of Section 7.3, for the combined doubly cyclic series plant, is modified for the open circuit/closed cycle plant. The work output from the gas-turbine plant of Fig. 7.3 is

$$W_H = (\eta_O)_H F, \tag{7.10}$$

$(\eta_O)_H$ is the (arbitrary) overall efficiency and F is the energy supplied in the fuel, $F = M_f[CV]_0$, where $[CV]_0$ is the enthalpy of combustion of the fuel of mass flow M_f. The work output from the steam cycle is

$$W_L = \eta_L Q_L, \tag{7.11}$$

in which η_L is the thermal efficiency of the lower (steam) cycle and Q_L is the heat transferred from the gas turbine exhaust.

Thus, the (arbitrary) overall efficiency of the whole plant is

$$(\eta_O)_{CP} = \frac{W_H + W_L}{F} = (\eta_O)_H + \frac{\eta_L Q_L}{F}. \tag{7.12}$$

But if combustion is adiabatic, then the steady flow energy equation for the open-circuit gas turbine (with exhaust of enthalpy $(H_P)_S$ leaving the HRSG and entering the exhaust stack with a temperature T_S greater than that of the atmosphere, T_0) is

$$H_{R0} = H_{PS} + W_H + Q_L, \tag{7.13}$$

---

[1] Note that in Fig. 7.3, the steam entropy is scaled by a factor $\mu = M_s/M_g$, obtained from the heat balance, $M_g(h_4 - h_6) = M_g \int_6^4 T ds_g = M_s(h_e - h_c) = M_s \int_c^e T ds_s$. Point c is then vertically under point 6 (but point b may not be precisely vertically below point S).

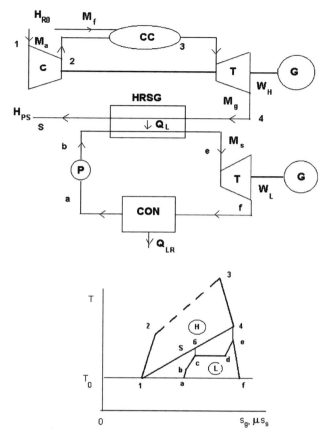

Fig. 7.3. Open circuit gas turbine/closed steam cycle combined plant (CCGT). No supplementary firing (after Ref. [1]).

so that

$$Q_L = H_{R0} - H_{PS} - W_H = [H_{R0} - H_{P0}] - [H_{PS} - H_{P0}] - W_H$$

$$= F - [H_{PS} - H_{P0}] - W_H, \tag{7.14}$$

where H_{P0} is the enthalpy of products leaving the calorimeter in a 'calorific value' experiment, after combustion of fuel M_f at temperature T_0.

The arbitrary overall efficiency of the combined plant (Eq. (7.12)) may then be written as

$$(\eta_O)_{CP} = (\eta_O)_H + \eta_L \left[1 - \frac{W_H}{F} - \frac{[H_{PS} - H_{P0}]}{F} \right]$$

$$= (\eta_O)_H + \eta_L - (\eta_O)_H \eta_L - \frac{\eta_L [H_{PS} - H_{P0}]}{F}, \tag{7.12a}$$

$$= (\eta_O)_H + (\eta_B)\eta_L - \eta_B(\eta_O)_H\eta_L, \tag{7.12b}$$

$$= (\eta_O)_H + (\eta_O)_L - (\eta_O)_H(\eta_O)_L, \tag{7.12c}$$

where

$$\eta_B = 1 - [(H_P)_S - (H_P)_0]/F[1 - (\eta_O)_H].$$

Expression (7.12a) for overall efficiency is similar to that for the combined doubly cyclic plant; the term $\eta_L[H_{PS} - H_{P0}]/F$ corresponds to the 'heat loss' term of Section 7.3. The extent of this reduction in overall efficiency depends on how much exhaust gases can be cooled and could theoretically be zero if they emerged from the HRSG at the (ambient) temperature of the reactants. In practice this is not possible, as corrosion may take place on the tubes of the HRSG if the dew point temperature of the exhaust gases is above the feed water temperature. We shall find that there may be little or no advantage in using feed heating in the steam cycle of the CCGT plant.

7.4.2. The integrated coal gasification combined cycle plant (IGCC)

A current development of the exhaust heated plant (unfired) is the integrated coal gasification combined cycle (IGCC) plant. One of the earliest of these IGCCs was the Cool Water pilot plant built by the General Electric company, using a Texaco gasifier. This complex plant is shown in Fig. 7.4, after Plumley [4]. The gas turbine, HRSG and steam turbine components were standard so it was the performance of the gasifier which was critical for new development and close integration between the gasifier and the HRSG was important.

In the plant, coal is ground and mixed with water to form a slurry and this is fed to the gasifier through a burner, in which partial combustion takes place with oxygen (supplied from a separate plant). During gasification the coal ash is melted into a slag, quenched with water and removed as a solid.

Following the high temperature reactions of coal and water with oxygen, the raw synthetic gas (syngas), consisting mainly of hydrogen and carbon monoxide (about 40% each by molal concentration) is water-cooled in radiant and convection coolers, generating saturated steam. The gas is then passed through a particulate scrubber, further cooled to near ambient temperature prior to sulphur removal, and then saturated to reduce the subsequent combustion temperature and NO_x production.

The syngas then enters the conventional exhaust heated CCGT plant, being burnt in the gas turbine combustion chamber with air from the compressor. The combustion gas supplies the gas turbine, driving the compressor and a generator, and then exhausts into the HRSG (unfired), which raises superheated steam. By-product steam from the gasifier coolers (some 40% of the total steam supply) is also superheated in the HRSG and the two streams of steam enter the steam turbine which drives its own generator.

Some 20 IGCC plants, in various forms, some with other gasifiers but most using oxygen, are now operating or are in the process of construction. Modifications of the IGCC plant to sequestrate the carbon dioxide produced will be discussed in Chapter 8.

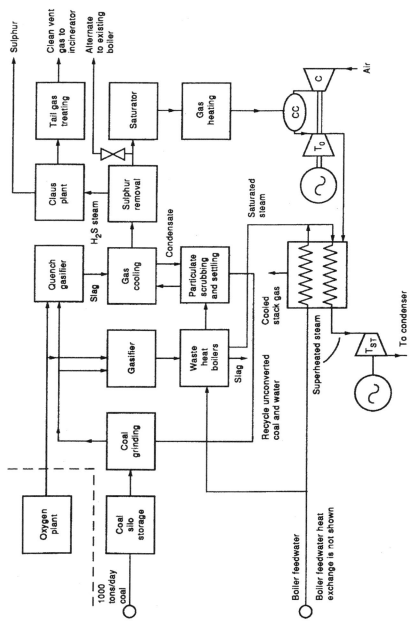

Fig. 7.4. Cool water IGCC plant (after Plumley [4]).

7.4.3. The exhaust heated (supplementary fired) CCGT

The exhaust gases from a gas turbine contain substantial amounts of excess air, since the main combustion process has to be diluted to reduce the combustion temperature to well below that which could be obtained in stoichiometric combustion, because of the metallurgical limits on the gas turbine operating temperature. This excess air enables supplementary firing of the exhaust to take place and higher steam temperatures may then be obtained in the HRSG.

The T, s diagram for a combined plant with supplementary firing is illustrated in Fig. 7.5 (again the steam entropy has been scaled). Introduction of regenerative feed heating of the water is of doubtful value, as will be discussed later. Supplementary heating generally lowers the overall efficiency of the combined plant. Essentially this is because a fraction of the total heat supplied is utilised to produce work in the lower cycle, of lower efficiency than that of the higher cycle.

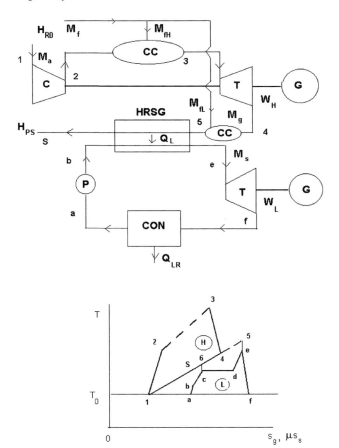

Fig. 7.5. Open circuit gas turbine/closed steam cycle combined plant (CCGT). With supplementary firing (after Ref. [1]).

For a mass flow of air (M_a) to the compressor of the gas turbine plant, a mass flow M_f of fuel (of specific enthalpy h_{f0}) is supplied to the *two* combustion chambers ($M_f = (M_f)_H + (M_f)_L$). The overall efficiency of the combined plant is then

$$(\eta_O)_{CP} = \frac{W_H + W_L}{M_f(CV)_0} = \frac{W_H + W_L}{[(M_f)_H + (M_f)_L][CV]_0}$$

$$= \frac{(\eta_O)_H}{\left[1 + \dfrac{(M_f)_L}{(M_f)_H}\right]} + \frac{\eta_L Q_L}{(M_f)_H[CV]_0\left[1 + \dfrac{(M_f)_L}{(M_f)_H}\right]}. \tag{7.15}$$

Eq. (7.15) may be written as

$$(\eta_O)_{CP} = \frac{(\eta_O)_H}{\left[1 + \dfrac{(M_f)_L}{(M_f)_H}\right]} + \frac{\eta_L}{\left[1 + \dfrac{(M_f)_L}{(M_f)_H}\right]}$$

$$\times \left\{\frac{(M_f)_L}{(M_f)_H} + [1 - (\eta_O)_H] - \frac{[H_{P'S} - H_{P'0}]}{(M_f)_H[CV]_0}\right\}, \tag{7.16}$$

where $H_{P'} = [M_a + (M_f)_H + (M_f)_L]h_{P'}$, and P' indicates products after supplementary combustion.

Eq. (7.16) may be written in terms of 'heating' quantities as

$$Q_H = (M_f)_H[CV]_0 \text{ and } Q_L = (M_f)_L[CV]_0$$

and a 'heat loss'

$$Q_{UN} = [M_a + (M_f)_H + (M_f)_L][(h_{P'})_S - (h_{P'})_0]$$

Then with $v_L = Q_L/(Q_L + Q_H)$ and $v_{UN} = Q_{UN}/(Q_L + Q_H)$, it follows that

$$(M_f)_L/(M_f)_H = v_L/(1 - v_L), \tag{7.17}$$

and

$$Q_{UN}/[(M_f)_H[CV]_0] = v_{UN}/(1 - v_L), \tag{7.18}$$

so that Eq. (7.16) becomes

$$(\eta_O)_{CP} = (\eta_O)_H + \eta_L - (\eta_O)_H\eta_L - \eta_L v_{UN} - (\eta_O)_H(1 - \eta_L)v_L. \tag{7.19}$$

7.5. The efficiency of an exhaust heated CCGT plant

The expression for the combined cycle efficiency

$$\eta = (\eta_O)_H + (\eta_O)_L[1 - (\eta_O)_H] \tag{7.20}$$

is always valid for CCGT exhaust heated (unfired) cycles. The parametric calculation of the efficiency of the upper open gas turbine plant ($\eta_O)_H$ is discussed in detail in Chapters 4 and 5. The overall efficiency of the lower steam cycle ($\eta_O)_L$ is the product of the lower thermal efficiency η_L and the 'boiler' efficiency of the HRSG, η_B.

Within the steam plant η_L depends on several factors:
- the boiler and condenser pressures;
- the turbine and boiler feed pump efficiencies;
- whether or not there is steam reheat;
- whether or not there is feed heating and whether the steam is raised in one, two or three stages.

On the other hand η_B depends on some of the following features of the gas turbine plant:
- the gas turbine final exit temperature;
- the specific heat capacity of the exhaust gases; and
- the allowable final stack temperature.

The interaction between the gas turbine plant and the steam cycle is complex, and has been the subject of much detailed work by many authors [5–8]. A detailed account of some of these parametric studies can be found in Ref. [1], and hence they are not discussed here. Instead, we first illustrate how the efficiency of the simplest CCGT plant may be calculated. Subsequently, we summarise the important features of the more complex combined cycles.

7.5.1. A parametric calculation

We describe a parametric 'point' calculation of the efficiency of a simple CCGT plant, firstly with no feed heating. It is supposed that the main parameters of the gas turbine upper plant (pressure ratio, maximum temperature, and component efficiencies) have been specified and its performance $(\eta_O)_H$ determined (Fig. 7.3 shows the T, s diagram for the two plants and the various state points).

For the steam plant, the condenser pressure, the turbine and pump efficiencies are also specified; there is also a single phase of water/steam heating, with no reheating. The feed pump work term for the relatively low pressure steam cycle is ignored, so that $h_b \approx h_a$. For the HRSG two temperature differences are prescribed:
(a)   the upper temperature difference, $\Delta T_{4e} = T_4 - T_e$; and
(b)   the 'pinch point' temperature difference, $\Delta T_{6c} = T_6 - T_c$.

With the gas temperature at turbine exit known (T_4), the top temperature in the steam cycle (T_e) is then obtained from (a). It is assumed that this is less than the prescribed maximum steam temperature.

If an evaporation temperature (p_c) is pre-selected as a parametric independent variable, then the temperatures and enthalpies at c and e are found; from (b) above the temperature T_6 is also determined. If there is no heat loss, the heat balance in the HRSG between gas states 4 and 6 is

$$M_g(h_4 - h_6) = M_s(h_e - h_c), \tag{7.21}$$

where M_g and M_s are the gas and steam flow rates, respectively. Thus, by knowing all the enthalpies the mass flow ratio $\mu = M_s/M_g$ can be obtained. As the entry water temperature T_b has been specified (as the condenser temperature approximately), a further application of the heat balance equation for the whole HRSG,

$$(h_4 - h_S) = \mu(h_e - h_b),$$ (7.22)

yields the enthalpy and temperature at the stack, (h_S, T_S).

Even for this simplest CCGT plant, iterations on such a calculation are required, with various values of p_c, in order to meet the requirements set on T_e, the steam turbine entry temperature, and T_S (the calculated value of T_S has to be such that the dewpoint temperature of the gas (T_{dp}) is below the economiser water entry temperature (T_b) and that may not be achievable). But with the ratio μ satisfactorily determined, the work output from the lower cycle W_L can be estimated and the combined plant efficiency obtained from

$$\eta_O = (W_H + W_L)/M_f[CV]_0,$$ (7.23)

as the fuel energy input to the higher cycle and its work output is already known.

This is essentially the approach adopted by Rufli [9] in a comprehensive set of calculations, but he assumed that the economiser entry water temperature T_b is raised above the condenser temperature by feed heating, which was specified for all his calculations. The T, s diagram is shown in Fig. 7.6; the feed pump work terms are neglected so that $h_a \approx h_{b'}$ and $h_{a'} \approx h_b$.

Knowing the turbine efficiency, an approximate condition line for the expansion through the steam turbine can be drawn (to state f' at pressure $p_{b'}$) and an estimate made of the steam enthalpy $h_{f'}$. If a fraction of the steam flow m_s is bled at this point then the heat balance for a direct heater raising the water from near the condenser temperature T_a to T_b is approximately

$$M_s(h_{f'} - h_b) = M_s(1 - m_s)(h_b - h_a),$$ (7.24)

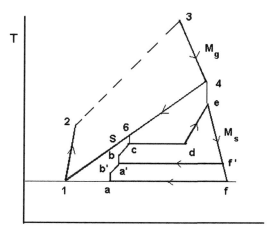

Fig. 7.6. CCGT plant with feed water heating by bled steam (after Ref. [1]).

and m_s can be determined. The work output from the steam cycle can then be obtained (allowing for the bleeding of the steam from the turbine) as

$$W_{L'} = M_s\{(h_e - h_{f'}) + (1 - m_s)(h_{f'} - h_f)\}, \tag{7.25}$$

where feed pump work terms have been neglected (the feed pumping will be split for the regenerative cycle with feed heating).

With the fuel energy input known from the calculation of the gas turbine plant performance, $F = M_f[CV]_0$, the combined plant efficiency is determined as

$$(\eta_O)_{CP} = (W_H + W_{L'})/F \tag{7.26}$$

The reason for using feed heating to set the entry feed water temperature at a level T_b above the condenser temperature T_a is that T_b must exceed the dewpoint temperature T_{dp} of the exhaust gases. If T_b is below T_{dp} then condensation may occur on the outside of the economiser tubes (the temperature of the metal on the outside of the tubes is virtually the same as the internal water temperature because of the high heat transfer on the water side). With $T_b > T_{dp}$ possible corrosion will be avoided.

Some of Rufli's calculations for $(\eta_O)_{CP}$, for a single boiler pressure p_c, are shown in Fig. 7.7a. There are two important features here:
(a) as expected, the overall CCGT efficiency increases markedly with gas turbine maximum temperature; and
(b) the optimum pressure ratio for maximum efficiency is low, relative to that for a simple CBT cycle. We return to this point below in Section 7.6.

Similarly comprehensive calculations were carried out by Cerri [10]:
(a) with and without feed heating, and
(b) with supplementary heating.

For (a), calculations showed that the presence of feed heating made little difference to the overall efficiency. Essentially, this is because although feed heating raises the thermal efficiency η_L, it leads to a higher value of T_S and hence a lower value of the boiler efficiency, η_B. The overall lower cycle efficiency $(\eta_O)_L = \eta_B \eta_L$ may be expected to change little in the expression for combined cycle efficiency $(\eta_O)_{CP}$, Eq. (7.12c). However, as pointed out before, feed heating can be used to ensure that T_b is higher than the dewpoint temperature of the exhaust gases, T_{dp}, to avoid corrosion of the economiser water tubes.

For (b), Cerri assumed that the supplementary 'heat supplied' was sufficient to give a maximum temperature equal to the assumed maximum steam entry temperature T_e. In general, it was shown that for the higher values of T_3 now used in CCGT plants there was little or no benefit on overall efficiency associated with supplementary heating.

Rufli also investigated whether raising the steam at two pressure levels showed any advantage. Typical results obtained by Rufli are also given in Fig. 7.7b. It can be seen that there is an increase of about 2–3% on overall efficiency resulting from two stages of heating rather than a single stage.

Results similar to the calculations of Rufli and Cerri have been obtained by many authors [5–8].

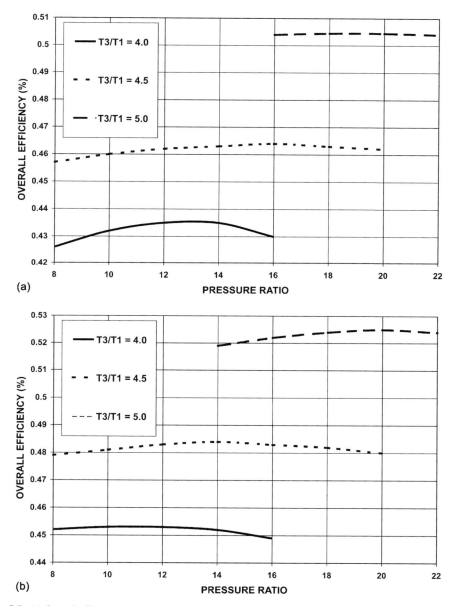

Fig. 7.7. (a) Overall efficiency of CCGT plant with feed water heating by bled steam and single pressure steam raising (after Rufli [9]). (b) Overall efficiency of CCGT plant with feed water heating with bled steam and dual pressure steam raising (after Rufli [9]).

7.5.2. Regenerative feed heating

For a comprehensive discussion on feed heating in a CCGT plant, readers may refer to Kehlhofer's excellent practical book on CCGTs [2]; a summary of this discussion is given below.

Kehlhofer takes the gas turbine as a 'given' plant and then concentrates on the optimisation of the steam plant. He discusses the question of the limitation on the stack and water entry temperatures in some detail, their interaction with the choice of p_c in a single pressure steam cycle, and the choice of two values of p_c in a dual pressure steam cycle. Considering the economiser of the HRSG he also argues that the dewpoint of the gases at exhaust from the HRSG must be less than the feed-water entry temperature; for sulphur free fuels the water dewpoint controls, whereas for fuels with sulphur a 'sulphuric acid' dewpoint (at a higher temperature) controls. Through these limitations on the exhaust gas temperature, the choice of fuel with or without sulphur content (distillate oil or natural gas, respectively) has a critical influence ab initio on the choice of the thermodynamic system.

For the simple single pressure system with feed heating, Kehlhofer first points out that the amount of steam production (M_s) is controlled by the pinch point condition if the steam pressure (p_c) is selected, as indicated earlier (Eq. (7.21)). However, with fuel oil containing sulphur, the feed-water temperature at entry to the HRSG is set quite high (T_b is about 130°C), so the heat that can be extracted from the exhaust gases beyond the pinch point $[M_s(h_c - h_b)]$ is limited. As shown by Rufli, the condensate can be brought up to T_b by a single stage of bled steam heating, in a direct contact heater, the steam tapping pressure being set approximately by the temperature T_b.

Kehlhofer then suggests that more heat can be extracted from the exhaust gases, even if there is a high limiting value of T_b (imposed by use of fuel oil with a high sulphur content). It is thermodynamically better to do this without regenerative feed heating, which leads to less work output from the steam turbine. *For a single pressure system with a pre-heating loop*, the extra heat is extracted from the exhaust gases by steam raised in a low pressure evaporator in the loop (as shown in Fig. 7.8, after Wunsch [11]). The evaporation temperature will be set by the 'sulphuric acid' dewpoint (and feed water entry temperature $T_b \approx 130$°C). The irreversibility involved in raising the feed water to temperature T_b is split between that arising from the heat transfer from gas to the evaporation (pre-heater) loop and that in the deaerator/feed heater. It is shown in Ref. [1] that the total irreversibility is just the same as that which would have occurred if the water had been heated from condenser temperature entirely in the HRSG. Thus, the simple method of calculation described at the beginning of Section 7.5.1 (with no feed water heating and $T_b \approx T_a$) is valid.

Kehlhofer explains that the pre-heating loop must be designed so that the heat extracted is sufficient to raise the temperature of the feed water flow from condenser temperature T_a to $T_{a'}$ (see Fig. 7.6). The available heat increases with live steam pressure (p_c), for selected $T_b(\approx T_a)$ and given gas turbine conditions, but the heat required to preheat the feed water is set by ($T_{a'} - T_a$). The live steam pressure is thus determined from the heat balance in the pre-heater if the heating of the feed water by bled steam is to be avoided; but the optimum (low) live steam pressure may not be achievable because of the requirement set by this heat balance.

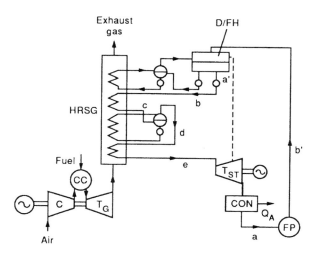

Fig. 7.8. Single pressure steam cycle system with LP evaporator in a pre-heating loop, as alternative to feed heating (after Wunsch [11]).

Kehlhofer regards the *two pressure system* as a natural extension of the single pressure cycle with a low pressure evaporator acting as a pre-heater. Under some conditions more steam could be produced in the LP evaporator than is required to pre-heat the feed water and this can be used by admitting it to the turbine at a low pressure. For a fuel with high sulphur content (requiring high feed water temperature (T_b) at entry to the HRSG), *a dual pressure system with no low pressure water economiser* may have two regenerative surface feed heaters and a pre-heating loop. For a sulphur free fuel (with a lower T_b), *a dual pressure system with a low pressure economiser* may have a single-stage deaerator/direct contact feed heater using bled steam.

7.6. The optimum pressure ratio for a CCGT plant

Rufli's calculations (Fig. 7.7a, b), indicated that the optimum pressure ratio for a CCGT plant is relatively low compared with that of a simple gas turbine (CBT) plant. In both cases, the optimum pressure ratio increases with maximum temperature. Davidson and Keeley [6] have given a comparative plot of the efficiencies of the two plants (Fig. 7.9), showing that the optimum pressure ratio for a CCGT plant is about the same as that giving maximum specific work for a CBT plant.

The reason for this choice of low pressure ratio is illustrated by an approximate analysis [12], which extends the graphical method of calculating gas turbine performance described in Chapter 3. If the gas turbine higher plant is assumed to operate on an air standard cycle (i.e. the working fluid is a perfect gas with a constant ratio of specific heats, γ), then the compressor work, the turbine work, the net work output and the heat supplied may be written as

$$\text{NDCW} = w'_C = (x - 1)/\eta_C(\theta - 1), \tag{7.27}$$

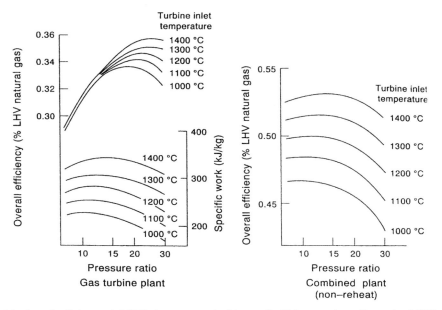

Fig. 7.9. Overall efficiency of CCGT plant compared with overall efficiency and specific work of CBT plant
(after Davidson and Keeley [6]).

$$\text{NDTW} = w'_T = \eta_T \theta (x - 1)/x(\theta - 1), \tag{7.28}$$

$$\text{NDNW} = w'_H = w'_T - w'_C, \tag{7.29}$$

$$\text{NDHT} = q'_H = (1 - w'_C), \tag{7.30}$$

respectively, where the primes indicate that all have been made non-dimensional by dividing by the product of the gas flow rate and $c_p(T_3 - T_1)$. These quantities are plotted against $x = r^{(\gamma-1)/\gamma}$ in Fig. 7.10, constant values being assumed for $\theta = (T_3/T_1) = 5.0$ and compressor and turbine efficiencies ($\eta_C = 0.9$, $\eta_T = 0.889$, $\eta_C \eta_T = 0.8$).

Timmermans [13] suggested that the steam turbine work output (per unit gas flow in the higher plant) is given approximately by

$$w_L = K c_p(T_4 - T_6) \tag{7.31}$$

where T_4 is the temperature at gas turbine exit, T_6 is the temperature in the HRSG at the lower pinch point and K is a constant (about 4.0). The (non-dimensional) steam turbine work can then be written as

$$\text{NDSTW} = w'_L = K(T_4 - T_6)/(T_3 - T_1) \tag{7.32}$$

and the total (non-dimensional) work output from the combined plant becomes

$$\text{NDCPW} = w'_{CP} = (1 - K)w'_H + K q'_H - k \tag{7.33}$$

where $k = K[(T_6/T_1) - 1]/(\theta - 1)$ is a small quantity and for an approximate analysis may be taken as constant ($k \approx 0.06$).

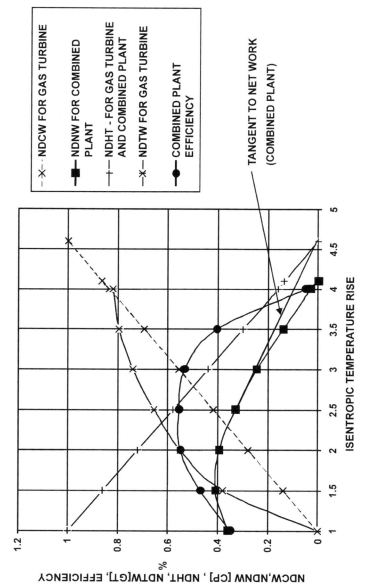

Fig. 7.10. Graphical plot showing determination of pressure ratio for maximum efficiency of CCGT plant (after Ref. [11]).

It can be seen from Fig. 7.10 that the curve for w'_{CP} lies above that for w'_H. As for the gas turbine cycle the pressure ratio for maximum efficiency in the combined plant may be obtained by drawing a tangent to the work output curve from a point on the x-axis where $x = 1 + \eta_c(\theta - 1)$, i.e. $x = 4.6$ in the example. The optimum pressure ratio for the combined plant ($r \approx 18$) is less than that for the gas turbine alone ($r \approx 30$) although it is still greater than the pressure ratio which gives maximum specific work in the higher plant ($r \approx 11$). However, the efficiency η_{CP} varies little with r about the optimum point.

It may also be noted that by differentiating Eq. (7.9) with respect to r (or x), and putting the differential equal to zero for the maximum efficiency, it follows that

$$\frac{\partial(\eta_O)_H}{\partial x} = -\frac{(1 - (\eta_O)_H)}{[1 - (\eta_O)_L]}\frac{\partial(\eta_O)_L}{\partial x} \tag{7.34}$$

and

$$\frac{\partial(\eta_O)_H}{\partial x} \approx -\frac{\partial(\eta_O)_L}{\partial x}, \tag{7.35}$$

since $(\eta_O)_H$ and $(\eta_O)_L$ are little different in most cases. Hence, the maximum combined cycle efficiency $(\eta_O)_{CP}$ occurs when the efficiency of the higher cycle increases with r at about the same rate as the lower cycle decreases. Clearly, this will be at a pressure ratio less than that at which the higher cycle reaches peak efficiency, and when the lower cycle efficiency is decreasing because of the dropping gas turbine exit temperature.

This approach was well illustrated by Briesch et al. [14], who showed separate plots of $(\eta_O)_H$, $(\eta_O)_L$ and $(\eta_O)_{CP}$ against pressure ratio for a given T_{max} and T_{min} (Fig. 7.11), illustrating the validity of Eq. (7.35). But note that the limiting allowable steam turbine entry temperature also influences the choice of pressure ratio in the gas turbine cycle.

7.7. Reheating in the upper gas turbine cycle

The case for supplementary heating at the gas turbine exhaust has already been considered; Cerri [10] showed that it leads to lower overall combined plant efficiency, except at low maximum temperature. Although there is a case for supplementary heating giving higher specific work, the modern CCGT plant with its higher gas turbine inlet temperature does not in general use supplementary heating. However, there is an argument for reheating in the gas turbine itself (i.e. between HP and LP turbines), which should lead to higher mean temperatures of supply and high overall efficiency.

Rice [15] made a comprehensive study of the reheated gas turbine combined plant. He first analysed the higher (gas turbine) plant with reheat, obtaining $(\eta_O)_H$, turbine exit temperature, and power turbine expansion ratio, all as functions of plant overall pressure ratio and firing temperatures in the main and reheat burners. (The optimum power turbine expansion ratio is little different from the square root of the overall pressure ratio.) He then pre-selected the steam cycle conditions rather than undertaking a full optimisation.

Rice argued that a high temperature at entry to the HRSG (resulting from reheat in the gas turbine plant) leads via the pinch point restriction to a lower exhaust stack temperature and 'heat loss', in comparison with an HRSG receiving gas at a lower temperature from

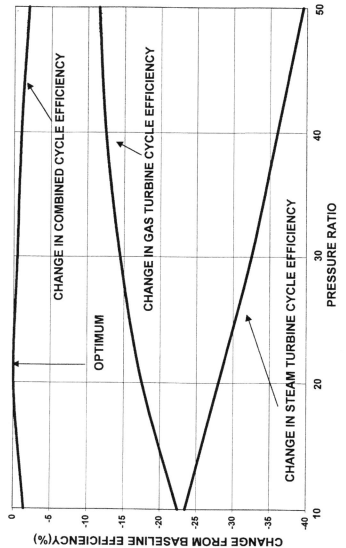

Fig. 7.11. Change in overall efficiency with pressure ratio for CCGT plant and component plants (after Briesch et al. [14]).

a simple gas turbine plant. But there are additional complications, of higher irreversibility in the HRSG (because of higher temperature differences), the possibility of regenerative feed heating and the limitation on the temperature of the water at entry to the HRSG economiser.

Rice found high CCGT efficiencies with gas turbine reheat at optimum pressure ratios even higher than those discussed above.

The latest ABB GT24/36 CCGT plant ([16], see also Ref. [1] for a brief description) employs reheating between the HP and LP turbines and a relatively high pressure ratio of 30. There are two thermodynamic features of this type of design. Firstly, the expansion through the larger pressure ratio, but taken in separate HP and LP turbines with reheating between them, means that the temperature leaving the LP turbine is not increased substantially in comparison with non-reheated plants (about 600°C, cf. 530–500°C); and secondly that the pressure ratio for maximum $(\eta_O)_{CP}$ becomes closer to that for the maximum efficiency in the higher plant alone.

An extension of the approximate analysis of Section 7.6 suggests that the pressure ratio for both the combined and higher level plants, for the example given there, should be about 48 which is higher than that used in the ABB plant (about 30).

Most modern CCGT plants use open air cooling in the front part of the gas turbine. An exception is the GE MS9001H plant which utilises the existence of the lower steam plant to introduce steam cooling of the gas turbine. This reduces the difference between the combustion temperature T_{cot} and the rotor inlet temperature T_{rit} The effect of this on the overall combined plant efficiency is discussed in Ref. [1] where it is suggested that any advantage is small.

7.8. Discussion and conclusions

(a) It has been shown that the CCGT plant achieves a much higher overall efficiency than the simple CBT plant, but the maximum efficiency is achieved at a substantially lower pressure ratio than that giving optimum conditions in the latter plant.

(b) With modern gas turbine inlet temperatures there is no advantage in supplementary heating. However, reheating in the gas turbine may give high efficiency, but at a higher optimum pressure ratio.

(c) Irreversibility in the HRSG may be reduced by introducing dual pressure level steam raising. This may increase the overall efficiency by about 2–3%, but going to triple pressure levels adds relatively little further gain.

(d) The introduction of feed heating into the steam cycle of a CCGT plant is a complex matter and the following points are relevant.

    (i) The simplest recuperative plant, with no regenerative feed heating and all the feed water heated directly in the HRSG may not be feasible because of the limits that have to be placed on the temperature T_b of the feed water entering the HRSG (in order to avoid corrosion of the metal surfaces). However, a thermodynamic performance the same as this simplest plant (no regenerative feed heating) can be achieved by extracting from the exhaust gases the heat required to raise the feed water from condenser temperature (T_a) to the prescribed minimum temperature (T_b) by a pre-heating loop. This modified plant should give higher efficiency than a plant which uses bled steam to heat the feed water to T_b.

(ii) The two pressure steam cycle using fuel containing sulphur may require the introduction of substantial feed heating and a pre-heating loop. In practice, this appears to provide the highest overall efficiency of the combined plant [2].

References

[1] Horlock, J.H. (2002), Combined Power Plants, 2nd edn, Krieger, Melbourne, USA.

[2] Kehlhofer, R. (1991), Combined Cycle Gas and Steam Turbine Power Plants, Fairmont Press, Lilburn, GA.

[3] Seippel, C. and Bereuter, R. (1960), The theory of combined steam and gas turbine installations, Brown Boveri Review 47, 783–799.

[4] Plumley, D.R. (1985), Cool water coal gasification 1—A progress report, ASME J. Engng Power Gas Turbines 107(4), 856–860.

[5] Bolland, O.A. (1991), Comparative evaluation of advanced combined cycle alternatives, ASME J. Engng Gas Turbines Power 113(2), 190–195.

[6] Davidson, B.J. and Keeley, K.R. (1991), The thermodynamics of practical combined cycles, Proc. Instn. Mech. Engrs. Conference on Combined Cycle Gas Turbines, 28–50.

[7] Finckh, H.H. and Pfost, H. (1992), Development potential of combined cycle [GUD] power plants with and without supplementary firing, ASME J. Engng Power Gas Turbines 114(4), 653–659.

[8] Jerica, H. and Hoeller, F. (1991), Combined cycle enhancement, ASME J. Engng Gas Turbines Power 113(2), 198–202.

[9] Rufli, P.A. (1987), A systematic analysis of the combined gas–steam cycle, Proc. ASME COGEN—Turbo I, 135–146.

[10] Cerri, G. (1987), Parametric analysis of combined cycles, ASME J. Engng Gas Turbines Power 109(1), 46–55.

[11] Wunsch, A. (1978), Combined gas/steam turbine power stations—the present state of progress and future developments, Brown Boveri Review 65(10), 646–655.

[12] Horlock, J.H. (1995), The optimum pressure ratio for a CCGT plant, Proc. Instn. Mech. Engrs. 209, 259–264.

[13] Timmermans, A.R.J. (1978), Combined Cycles and Their Possibilities. In Von Karman Institute for Fluid Dynamics, Lecture Series 6, Vol. 1.

[14] Briesch, M.S., Bannister, R.L., Dinkunchak, I.S. and Huber, D.J. (1995), A combined cycle designed to achieve greater than 60% efficiency, ASME J. Engng Gas Turbines Power 117(1), 734–741.

[15] Rice, I.G. (1980/1991), The combined reheat gas turbine/steam turbine cycle, ASME J. Engng Gas Turbines Power, 102, 1, Part I, 35–41, Part II, 42–49; 13, 2, 198–202.

[16] ABB Power Generation (1997), The GT24/26 gas turbines, ABB Brochure PGT2186.

Chapter 8

NOVEL GAS TURBINE CYCLES

8.1. Introduction

In the previous chapters, we have been concerned mainly with the thermodynamics of 'standard' gas turbine cycles, in a variety of forms. In this chapter, we consider some novel types of gas turbine cycles recently proposed, most of which have not yet been built.

So far, we have focussed on the achievement of maximum thermal efficiency and maximum specific work in power producing plants (or maximum energy utilisation and fuel savings in cogeneration plants). Practical gas turbines built up to the present time have been mainly based on those cycles already described, with designers seeking higher efficiency through

(a) advancing the basic thermodynamic parameters (such as turbomachinery polytropic efficiency, turbine inlet temperature, and compressor pressure ratio);

(b) use of better materials able to withstand higher temperatures; and

(c) introducing additional features, such as recuperation, intercooling, reheating, water injection, etc.

But for power station applications, the thermal efficiency is not the only measure of the performance of a plant. While a new type of plant may involve some reduction in running costs due to improved thermal efficiency, it may also involve additional capital costs. The cost of electricity produced is the crucial criterion within the overall economics, and this depends not only on the thermal efficiency and capital costs, but also on the price of fuel, operational and maintenance costs, and the taxes imposed. Yet another factor, which has recently become important, is the production by gas turbine plants of greenhouse gases (mainly carbon dioxide) which contribute to global warming. Many countries are now considering the imposition of a special tax on the amount of CO_2 produced by a power plant, and this may adversely affect the economics. So consideration of a new plant in future will involve not only the factors listed above but also the amount of CO_2 produced per unit of electricity together with the extra taxes that may have to be paid.

A brief and simplified description of how electricity price may be determined is given in Appendix B, giving some comparisons between different basic plants. We also describe there how the economics of a new plant may be affected by the imposition of an extra carbon tax associated with the amount of carbon dioxide produced.

Thus there are now three objectives for the plant designer:

(i)   high efficiency
(ii) low capital cost; and
(iii) a low quantity of carbon dioxide discharged to the atmosphere (either intrinsically low production or sequestration, liquefaction, removal and disposal of that produced by the plant).
In some of the plants proposed these objectives are attained simultaneously.

8.2. Classification of gas-fired plants using novel cycles

Against this background of the changed economics of plant performance, we consider some of the many new gas turbine plants that have been proposed over the past few years. In this section, we first formulate a list and classify these plants (and the 'cycles' on which they are based), as in Tables 8.1A–D, noting that most but not all use natural gas as a fuel.

8.2.1. Plants (A) with addition of equipment to remove the carbon dioxide produced in combustion

These cycles allow sequestration and disposal of CO_2 as a liquid, rather than allowing it to enter the atmosphere. They involve the introduction of additional equipment for the CO_2 removal but little or no modification of the basic CBT or CBTX plant itself.

Three such plants are:

A1   An open CCGT plant with 'end of pipe' removal of CO_2;

A2   A 'semi-closed' CCGT plant, involving recirculation of part of the exhaust gases, enabling the CO_2 to be separated more easily; and

A3   A 'semi-closed' CBTX plant, involving recirculation of part of the exhaust gases downstream of the heat exchanger, which also enables the CO_2 to be separated more easily.

Use of similar removal equipment in a simple CBT cycle is also possible but the exhaust gas from the turbine would require cooling before sequestration.

Table 8.1A
Cycles A with addition of CO_2 equipment

Description	Type	Special features	Fuel/oxidant	CO_2 removal	Comment
A1 'End of pipe' CO_2 removal	Open/CCGT	–	Natural gas/air	LP (chemical)	Simple CO_2 removal, but large CO_2 plant
A2 Semi-closed, CO_2 removal	SC/CCGT	–	Natural gas/air	LP (chemical)	Simple CO_2 removal, smaller CO_2 plant
A3 Recuperative, CO_2 removal	SC/CBTX	Recuperator	Natural gas/air	LP (chemical)	Simple CO_2 removal

Table 8.1B
Cycles B with combustion modification (fuel)

Description	Type	Special features	Fuel/oxidant	CO$_2$ removal	Comment
B1 Steam/TCR	Open/CBT	CH$_4$/steam reforming	Natural gas/air	None	Attractive simplicity and efficiency
B2 Steam/TCR plus water shift reactions	Open/CCGT	CH$_4$/steam reforming	Natural gas/air	LP (chemical)	Increased complexity
B3 FG/TCR	SC/CBT	CH$_4$/steam reforming	Natural gas/air	None	Little efficiency gain

8.2.2. *Plants (B) with modification of the fuel in combustion—chemically reformed gas turbine (CRGT) cycles*

These cycles involve modification of the combustion process, and employ thermo-chemical recuperation (TCR) to produce a fuel of higher hydrogen content. Three simple CRGTs are:

B1 the steam/TCR plant—mixing the fuel with steam raised in a heat recovery steam generator;

B2 the steam/TCR plant, with additional equipment for CO$_2$ removal;

B3 the Flue Gas/TCR cycle—mixing the fuel with partially recirculated exhaust gases containing water vapour.

In these CRGT plants, efficiency increase is obtained mainly through the abstraction of more heat from the exhaust gases rather than reduction in combustion irreversibility.

8.2.3. *Plants (C) using non-carbon fuel (hydrogen)*

Obviously the availability of a non-carbon fuel, usually hydrogen, would obviate the need for carbon dioxide extraction and disposal, and a plant with combustion of such a fuel becomes a simple solution (Cycle C1, a hydrogen burning CBT plant, and Cycles C2 and C3, hydrogen burning CCGT plants).

Table 8.1C
Cycles C with combustion using non-carbon fuel

Description	Type	Special features	Fuel/oxidant	CO$_2$ removal	Comment
C1 Hydrogen or hydrogen/nitrogen	Open/CBT	None	Hydrogen/air	None	Nitrogen compression required
C2 Hydrogen CCGT	Closed upper cycle CCGT	None	Hydrogen/air	None	High efficiency
C3 Rankine type double steam cycle	Closed upper cycle	None	Hydrogen/air	None	High efficiency

All depend on hydrogen availability.

Table 8.1D
Cycles D with combustion modification (oxidant)

Description	Type	Special features	Fuel/oxidant	CO$_2$ removal	Comment
D1 Partial oxidation CBT	Open/CBT	PO/steam reforming	Natural gas/air	None	Efficiency gain via reheat
D2 Partial oxidation CCGT	Open/CCGT	PO/steam reforming	Natural gas/air	HP (physical absorption)	Efficiency gain via reheat
D3 Partial oxidation	SC/CICBTBTX	Multi-PO, steam reforming	Natural gas/air	None	Very high efficiency complex
D4 Full oxidation CBT	SC/CBT	None	Natural gas/oxygen	LP liquid extraction	Easy CO$_2$ removal
D5 Full oxidation CBT	SC/CBT	None	Natural gas/oxygen	HP (physical absorption)	–
D6 Matiant complex cycle	Almost closed	Reheat, recuperator vapour/liquid compression train	Natural gas/oxygen	HP (liquid extraction)	Complex but high efficiency

8.2.4. *Plants (D) with modification of the oxidant in combustion*

In conventional cycles, combustion is the major source of irreversibility, leading to reduction in thermal efficiency. Some novel plants involve partial oxidation (PO) of the fuel in two or more stages, with the temperature increased before each stage of combustion, and the combustion irreversibility consequently reduced. In other plants full oxidation is employed which makes CO_2 removal easier.

Six cycles with oxidant modification are listed as
D1   the simple PO open CBT cycle—involving staged combustion of the fuel;
D2   the PO open CCGT cycle—involving staged combustion of the fuel and low pressure CO_2 removal;
D3   the semi-closed CICBTBTX cycle—involving staged partial combustion of the fuel, intercooling, recuperation and low pressure CO_2 removal;
D4   the 'semi-closed' CBT or CCGT plant with full oxidation—oxygen supplied to the combustion chamber instead of air, with CO_2 removal at low pressure level;
D5   the 'semi-closed' CBT plant with full oxidation—oxygen supplied to the combustion chamber instead of air, with CO_2 removal at high pressure level;
D6   the Matiant cycle—an almost closed CICBTBTX cycle using full oxidation and full CO_2 removal.

8.2.5. *Outline of discussion of novel cycles*

Below we describe
(i)   the additional equipment that is required for plants with CO_2 sequestration and liquefaction, at high or low pressure (in Section 8.3);
(ii) the concept of the 'semi-closed' cycle which features in some of the proposed plants (in Section 8.4); and
(iii) the various chemical reactions involved in combustion modification, through chemical recuperation, PO, etc. (in Section 8.5).

We then discuss in more detail the individual cycles listed above (in Section 8.6).

We also give calculations of the performance of some of these various gas turbine plants. Comparison between such calculations is often difficult, even 'spot' calculations at a single condition with state points specified in the cycle, because of the thermodynamic assumptions that have to be made (e.g. how closely conditions in a chemical reformer approach equilibrium). Performance calculations by different inventors/authors are also dependent upon assumed levels of component performance such as turbomachinery polytropic efficiency, required turbine cooling air flows and heat exchanger effectiveness; if these are not identical in the cases compared then such comparisons of overall performance become invalid. However, we attempt to provide some performance calculations where appropriate in the rest of the chapter.

Finally, in Section 8.7, we describe some modifications of the integrated gasification combined cycle (IGCC) which enable CO_2 to be removed (Cycles E).

8.3. CO_2 removal equipment

There are two main schemes proposed for sequestration of carbon dioxide. The first (referred to as a *chemical* absorption process), suitable for use at low pressures and temperatures, is usually adopted where the CO_2 is to be removed from exhaust flue gases. The second (usually referred to as a *physical* absorption process), for use at higher pressures, is recommended for separation of the CO_2 in syngas obtained from conversion of fuel.

8.3.1. The chemical absorption process

Fig. 8.1 shows a diagram of a chemical absorption process described by Chiesa and Consonni [1], for removal of CO_2 from the exhaust of a natural gas-fired combined cycle plant (in open or semi-closed versions). The process is favoured by low temperature which increases the CO_2 solubility, and ensures that the gas is free of contaminants which would impair the solvent properties.

Exhaust gas is fed to an absorber where the solvent (a blend of ethanol amines, mono-ethanolamine and di-ethanolamine) absorbs the carbon dioxide, and a CO_2 free stream is discharged to the atmosphere from the top of the absorption tower. Condensate is fed via a heat exchanger to a stripper from which the solvent is drained into a re-boiler (heat is supplied by steam fed from the HRSG of the combined cycle). Carbon dioxide and water leave the top of the stripper, passing through a cooler and separator, from which water is drained. Gaseous CO_2 leaves the top of the separator to enter an intercooled compressor; the compressed CO_2 is also aftercooled, and liquid carbon dioxide is discharged ready for disposal.

The negative aspects of the system on the combined cycle efficiency lie in the steam consumption for the stripping process, and the extra work inputs, to the CO_2 compressor and to the fans required to circulate the gases, through a system with non-negligible pressure losses. Corti and Manfrida [2] have considered in some detail the losses involved and argue that by careful optimisation of the composition of the amines blend in the solvent (50% di-ethanolamine in the aqueous solution containing the amine blends), the heat required for regeneration of the scrubbing solution can be limited. They have also drawn attention to the advantages of recovering combustion generated water into the lower steam cycle.

8.3.2. The physical absorption process

Fig. 8.2 shows a diagram of the physical absorption process suggested by Chiesa and Consonni [3] for an IGCC plant, with the absorption taking place from the syngas after its discharge at high pressure from the gasification and H_2S cleansing process. The CO_2 fed to the absorber is of a high concentration and flows upward, counter current to the CO_2 lean solvent (Selexol is proposed, which is soluble in CO_2 but not in nitrogen).

The CO_2 rich solvent is drained from the bottom of the tower, and led first to a hydraulic turbo-expander and then to four flash drums connected in series, where CO_2 is de-absorbed as the pressure is lowered. Lean solvent is pumped back to the top of the absorber tower

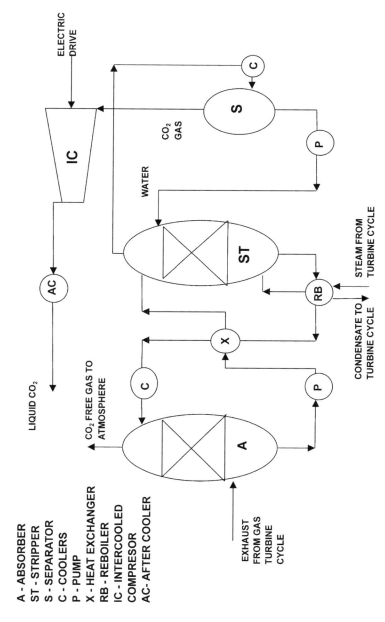

A - ABSORBER
ST - STRIPPER
S - SEPARATOR
C - COOLERS
P - PUMP
X - HEAT EXCHANGER
RB - REBOILER
IC - INTERCOOLED
COMPRESOR
AC- AFTER COOLER

Fig. 8.1. The chemical absorption process (after Chiesa and Consonni [1]).

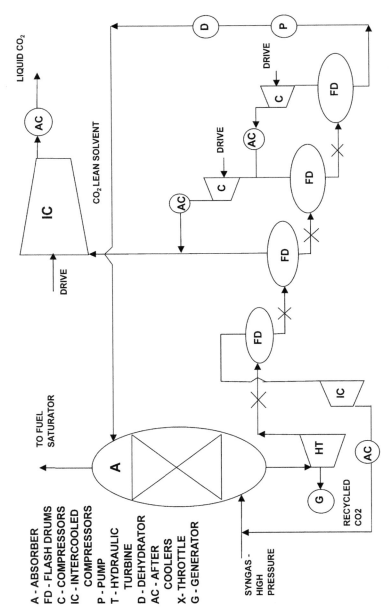

Fig. 8.2. The physical absorption process (after Chiesa and Consonni [3]).

from the last drum; carbon dioxide is collected from the other drums and compressed and intercooled for final discharge.

Manfrida [4] argues that the heat demand and the substantial power loss associated with 'presssure-swing' physical absorption makes it less attractive than chemical absorption, even for high pressure sequestration. The expansion work in the former is difficult to recover as several expanders are needed.

8.4. Semi-closure

Most of the novel cycles considered later in this chapter involve 'semi-closure', i.e. recirculation of some part of the exhaust gases into the compressor as indicated in the simplest example shown in Fig. 8.3. In effect, the exhaust products stream becomes an oxygen carrier.

Here, we first discuss whether such semi-closure (which is introduced so that CO_2 separation can be undertaken more easily) is likely to lead to higher or lower thermal efficiency, and in this discussion it is helpful to consider recirculation in relation to an air standard cycle (see Fig. 8.4). Fig. 8.4a shows a closed air standard cycle with unit air flow; Fig. 8.4b shows an open cycle similarly with unit air flow and an air heater rather than a combustion chamber. The cycles are identical in every respect except that in the former the turbine exhaust air from the turbine is cooled before it re-enters the compressor. In the latter, the turbine exhaust air is discharged to atmosphere and a fresh charge of air is taken in by the compressor. The quantities of heat supplied and the work output are the same for each of the two cycles, so that the thermal efficiencies are identical.

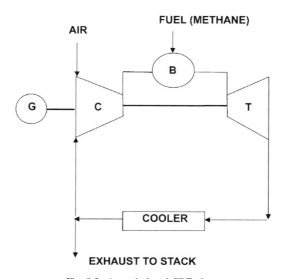

Fig. 8.3. A semi-closed CBT plant.

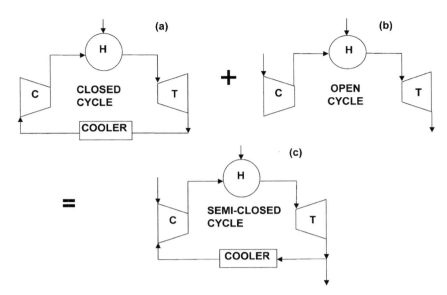

Fig. 8.4. Addition of a closed and an open cycle plant to form a semi-closed plant.

We can then add the two cycles together as shown in Fig. 8.4c, to form a semi-closed plant. There is double the flow through this new plant, double the heat supply and double the work output. Strictly, the total heat rejected is not doubled; half the turbine exhaust is now discharged to the atmosphere and half the heat rejected into a cooler before it is recirculated into the compressor. The thermal efficiency of this 'double' semi-closed plant is unchanged from that of the original closed cycle and the original open cycle. So there is apparently no thermodynamic advantage in semi-closure; it is undertaken for a different purpose.

A similar argument can be used for a fuelled semi-closed cycle, assuming that it can be regarded as the addition of an open CBT plant and a closed CHT cycle with identical working gas mass flow rates (and small fuel air ratios). Suppose the latter receives its heat supply from the combustion chamber of the former in which the open cycle combustion takes place. If the specific heats of air and products are little different, then the work output is doubled when the two plants are added together, but the fuel supply is also approximately doubled. The efficiency of the combined semi-closed plant is, therefore, approximately the same as that of the original open cycle plant.

8.5. The chemical reactions involved in various cycles

8.5.1. Complete combustion in a conventional open circuit plant

In the conventional gas turbine plant, a hydrocarbon fuel (e.g. methane CH_4) is burnt, usually with excess air, i.e. more air than is required for stoichiometric combustion.

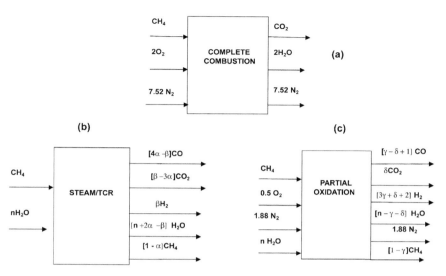

Fig. 8.5. Chemical reactions involved in various cycles.

Hence, all the carbon and hydrogen is used resulting in maximum formation of CO_2 and H_2O (complete combustion).

For a complete stoichiometric combustion of methane (Fig. 8.5a),

$$CH_4 + 2O_2 + 7.52N_2 \Rightarrow CO_2 + 2H_2O + 7.52N_2.$$

For combustion with say 200% excess air,

$$CH_4 + 6O_2 + 22.56N_2 \Rightarrow CO_2 + 2H_2O + 4O_2 + 22.56N_2.$$

Nitrogen is carried through the combustion unchanged and forms a large part of the 'carrying' gas for any unused oxygen. Supplementary combustion (or reheat) can then take place if more fuel is supplied to the products of primary combustion.

But in some of the novel cycles we shall consider that there may be
(i)   reforming of the fuel (into what is effectively a new fuel containing combustible CO and H_2); or
(ii) PO (i.e. incomplete combustion as insufficient air is available). We describe below the chemical reactions which may be involved in (i) and (ii).

8.5.2. *Thermo-chemical recuperation using steam (steam-TCR)*

The basic idea of using TCR in a gas turbine is usually to extract more heat from the turbine exhaust gases rather than to reduce substantially the irreversibility of combustion through chemical recuperation of the fuel. One method of TCR involves an overall reaction between the fuel, say methane (CH_4), and water vapour, usually produced in a heat recovery steam generator. The heat absorbed in the total process effectively increases the 'heating value' of the fuel before it is burnt in the combustion chamber. This does not necessarily mean that the calorific value is increased, but that the mass of the new fuel (syngas) may be increased so that the overall 'heating value' is also increased.

For the steam-TCR process, within a so-called 'Van't Hoff box' containing the total reaction process (Fig. 8.5b), there are two stages:

$$A: \quad CH_4 + H_2O \Leftrightarrow CO + 3H_2;$$

and

$$B: \quad CO + H_2O \Leftrightarrow CO_2 + H_2.$$

The so-called Boudouard reaction involving solid carbon is ignored here.

Stage A, the steam reforming reaction, is highly endothermic and stage B, usually known as the water gas shift reaction, is exothermic, so the overall reaction $(A + B)$ requires heat to be supplied. If this overall reaction is in equilibrium then the resulting mixture is made up of carbon monoxide, carbon dioxide, hydrogen, water vapour and remaining methane. Thus, if α moles of methane are converted (per mole supplied), and β moles of hydrogen are formed then the overall reaction may be written as

$$CH_4 + nH_2O \Rightarrow (4\alpha - \beta)CO + (\beta - 3\alpha)CO_2 + \beta H_2$$

$$+ (n + 2\alpha - \beta)H_2O + (1 - \alpha)CH_4,$$

where the total moles of the mixture are $N = (n + 1 + 2\alpha)$.

The net heat input that is required depends on the pressure p and the temperature T, and hence the equilibrium constants $K_{pA}(T)$ and $K_{pB}(T)$, respectively, which can be calculated as

$$K_{pA}(T) = \frac{(4\alpha - \beta)\beta^3 p^2}{(1 - \alpha)(n + 2\alpha - \beta)N^2}, \tag{8.1}$$

$$K_{pB}(T) = \frac{(\beta - 3\alpha)\beta}{(4\alpha - \beta)(n + 2\alpha - \beta)}. \tag{8.2}$$

With $(K_p)_A$ and $(K_p)_B$ known from tables of chemical data, then the various mole fractions, α, β, etc. may be determined if T and p are known.

Assuming that CH_4 and H_2O are supplied at T, the temperature at which TCR takes place, the heat required to produce the overall change (ΔH_{TCR}) is given by

$$[\Delta H]_{TCR} = (4\alpha - \beta)(h_{CO})_T + (\beta - 3\alpha)(h_{CO_2})_T + (\beta h_{H_2})_T + (2\alpha - \beta)(h_{H_2O})_T - \alpha(h_{CH_4})_T$$

$$= (4\alpha - \beta)[h_{CO} + 0.5h_{O_2} - h_{CO_2}]_T + \beta[h_{H_2} + 0.5h_{O_2} - h_{H_2O}]$$

$$- \alpha[h_{CH_4} + 2h_{O2} - h_{CO_2} - 2h_{H_2O}]_T$$

$$= (4\alpha - \beta)[\Delta H_{CO}]_T + \beta[\Delta H_{H_2}]_T - \alpha[\Delta H_{CH_4}]_T.$$

The 'heating value' of the resultant syngas mixture per mole of methane supplied, but now containing $(1 - \alpha)$ moles of CH_4, β moles of hydrogen and $(4\alpha - \beta)$ moles of carbon monoxide, is

$$[\Delta H]_{SYN} = \beta[\Delta H_{H_2}]_T + (1-\alpha)[\Delta H_{CH_4}]_T + (4\alpha - \beta)[\Delta H_{CO}]_T = [\Delta H]_{CH_4} + [\Delta H]_{TCR},$$

This is thus greater than the heating value of the original unit mole of methane supplied but is contained in a larger number of moles of syngas (N).

8.5.3. *Partial oxidation*

In the second chemical reaction to be considered, insufficient oxygen is supplied to the fuel for stoichiometric combustion (50%), but steam is also supplied (Fig. 8.5c). Now the chemical reactions involved in the partial combustion are:

$$A: \quad CH_4 + H_2O \Leftrightarrow CO + 3H_2,$$

the steam reforming reaction;

$$B: \quad CO + H_2O \Leftrightarrow CO_2 + H_2,$$

the water shift reaction; and

$$C: \quad CH_4 + 0.5O_2 \Leftrightarrow CO + 2H_2,$$

the PO reaction.

As in the steam/TCR analysis the Boudouard reaction is ignored here, together with direct methane decomposition.

The PO reaction, leading to five constituents, is now

$$2CH_4 + \tfrac{1}{2}O_2 + nH_2O$$

$$\Rightarrow (1-\gamma)CH_4 + \delta CO_2 + (\gamma - \delta + 1)CO + (3\gamma + \delta + 2)H_2 + (n - \gamma - \delta)H_2O$$

The solution then follows along the same lines as for TCR; if the temperature and pressure are known then γ, δ and the resulting mole fractions can be determined from the equilibrium constants. The temperature change between inlet and outlet is now likely to be higher than in the TCR reactions, so the determination of the K_ps as functions of a single mean temperature for the reaction is more difficult.

8.5.4. *Thermo-chemical recuperation using flue gases (flue gas/TCR)*

Another approach which has been suggested for thermo-chemical reforming can now be considered. It involves recirculation of exhaust gas from the turbine, which already contains some CO_2 and H_2O, to mix with the fuel in a reformer; the resulting syngas is then supplied to the main combustion chamber. The combustion process producing the flue gas is assumed to be virtually stoichiometric, with a small amount of excess air. The flue gas thus contains a small amount of oxygen and PO of the fuel (CH_4) may take place, together with the steam reforming and water shift reactions.

The 'Van't Hoff box' for this process will produce five components—carbon dioxide, carbon monoxide, water vapour and hydrogen, and unconverted methane. Again if the temperature T and pressure p are prescribed the mole fractions may be determined from the equilibrium constants, as described in the last section. The overall process is endothermic.

8.5.5. Combustion with recycled flue gas as a carrier

To complete the set of possible chemical reactions, consider the combustion of a fuel such as methane with a recirculated flue gas containing m moles of carbon dioxide, but assuming that water vapour has been removed from the recycling flue gas. If the additional air supply (n moles) is assumed to be sufficient for complete combustion, then

$$CH_4 + mCO_2 + nO_2 + 3.76nN_2 \Rightarrow (m + 1)CO_2 + 2H_2O + (n - 2)O_2 + 3.76nN_2.$$

From the products of combustion, CO_2 and $2H_2O$ may be removed subsequently within the recirculation cycle before the remaining mCO_2, reinforced with additional oxygen within the air supply, are fed back to the combustion chamber. Essentially, the complete combustion process described in Section 8.5.1 remains undisturbed by the 'carrying' recirculating flue gas.

8.6. Descriptions of cycles

With this background of how combustion may be modified we now study in some detail a number of novel cycles previously listed.

8.6.1. Cycles A with additional removal equipment for carbon dioxide sequestration

We consider first Cycles A of Table 8.1A and the associated Figs. 8.6–8.8. These are cycles in which the major objective is to separate or sequestrate some or all of the carbon dioxide produced, and to store or dispose it. This can be achieved either by direct removal of the CO_2 from the combustion gases with little or no modification to the existing plant; or by modest restructuring or alteration of the conventional power cycle so that the carbon dioxide can be removed more easily.

8.6.1.1. Direct removal of CO_2 from an existing plant

Fig. 8.6 shows an example of the first type of plant having an 'end of pipe' solution in which the CO_2 is removed from the exhaust of a standard CCGT plant, in an additional chemical absorption plant (Cycle A1). The products of combustion downstream of the HRSG (usually oxygen rich) are scrubbed by aqueous or organic based mixtures of amines. CO_2 in the exhaust gases is first absorbed and rich CO_2 liquid is then pumped to the stripper. The exhaust from the stripper is separated into water and gaseous CO_2, which is then compressed, intercooled and aftercooled before disposal as liquid CO_2 at high pressure and atmospheric temperature. A reasonably CO_2 free stream is passed to the stack and hence to the atmosphere.

Chiesa and Consonni [1] presented a detailed analysis of this type of plant. They found that the *net* efficiency of the plant dropped by about 5.5% below that of a basic CCGT plant

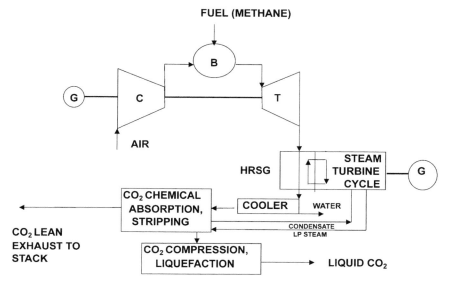

Fig. 8.6. Cycle A1. Direct removal of CO_2 from an existing plant (after Chiesa and Consonni [1]).

with some 56% efficiency, through addition of the absorption equipment. They also performed a detailed estimate of the extra capital cost, and found that the cost of electricity increased by some 40%, from 3.6 c/kWh for the basic plant to 5 c/kWh, due to the combined effect of lower efficiency and higher capital cost.

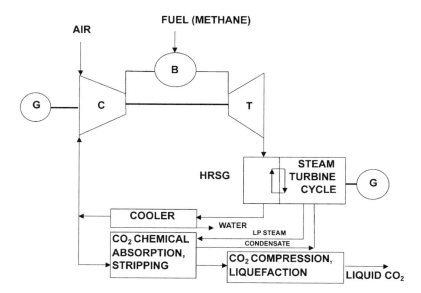

Fig. 8.7. Cycle A2. Semi-closed plant plus CO_2 removal (after Chiesa and Consonni [1]).

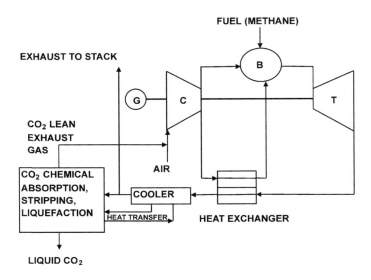

Fig. 8.8. Cycle A3. Semi-closed recuperative plant with CO_2 removal (after Manfrida [4]).

8.6.1.2. Modifications of the cycles of conventional plants using the semi-closed gas turbine cycle concept

Fig. 8.7 shows a second example (Cycle A2) of carbon dioxide removal by chemical absorption from a CCGT plant, but one in which the semi-closed concept is introduced—exhaust gas leaving the HRSG is partially recirculated. This reduces the flow rate of the gas to be treated in the removal plant, so that less steam is required in the stripper and the extra equipment to be installed is smaller and cheaper. This is also due to the better removal efficiency achievable—for equal reactants flow rate—when the volumetric fraction of CO_2 in the exhaust gas is raised from the 4–6% value typical of open cycle gas turbines to about 12% achievable with semi-closed operation.

Chiesa and Consonni [1] gave another detailed analysis for this plant in comparison with Cycle A1. They found that the efficiency dropped by 5% from that of the basic CCGT plant; this is somewhat surprising as the absorption plant is smaller than that for Cycle A1 and it might have been expected that the penalty on efficiency of introducing the absorption plant would have been much less than that of Cycle A1. With this calculated efficiency and a detailed estimate of capital cost, the price of electricity was virtually the same as that of Cycle A1, i.e. 40% greater than that of the basic CCGT plant.

Corti and Manfrida [2] have also done detailed calculations of the performance of plant A2. They drew attention to the need to optimise the amines blend (including species such as di-ethanolamine and mono-ethanolamine) in the absorption process, if a removal efficiency of 80% is to be achieved and in order to reduce the heat required for regenerating the scrubbing solution. Their initial estimates of the penalty on efficiency are comparable to those of Chiesa and Consonni (about 6% compared with the basic CCGT plant) but they emphasise that recirculation of water from the scrubbing process to intercool and aftercool the compression in the gas turbine cycle can restore about half the loss in thermal efficiency. After a very careful optimisation, and by including amine regeneration, Corti and Manfrida estimated the cost of electricity generated by this plant, including CO_2 disposal, to be about 4.7 c/kWh. This is slightly less than the estimate of Chiesa and Consonni who based their calculations on different sources.

Fig. 8.8 shows yet another example (Cycle A3) of the use of the semi-closed cycle concept, suggested by Manfrida [4], in which a recuperative CBTX plant is modified. Now the exhaust gas from the gas turbine is cooled in a heat exchanger (rather than the HRSG of a CCGT plant). It then enters the chemical absorption plant where some CO_2 is sequestrated and liquefied before disposal. The remainder of the exhaust gas is recirculated into compressor inlet after additional cooling. Manfrida finds slightly lower efficiency in the plant A3 compared with plant A2, but argues that it may prove simpler and more economic than the semi-closed IGCC plant.

8.6.2. *Cycles B with modification of the fuel in combustion through thermo-chemical recuperation [TCR]*

We consider next the cycles B of Table 8.1B and the associated Figs. 8.9–8.12; these cycles involve modification of the fuel used in the combustion process by TCR. There are two basic types of chemically recuperated gas turbine (CRGT) cycle:

(i) recuperative 'STIG type' cycles (B1, B2) in which the exhaust gas is used to raise steam in an HRSG, which is not then fed directly to the combustion chamber but first mixed with the fuel in a chemical reactor or reformer, the process described in Section 8.5.2 (in practice, the HRSG and the reformer may be combined in a single unit to form the syngas fuel);

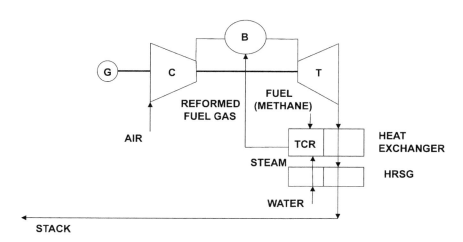

Fig. 8.9. Cycle B1. Chemically recuperated cycle with steam reforming.

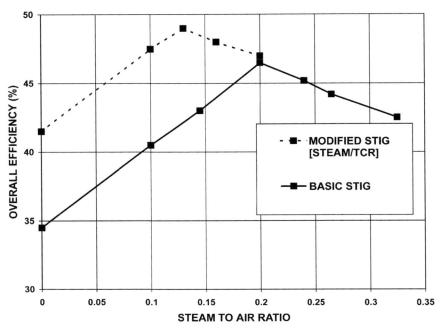

Fig. 8.10. Overall efficiencies of a steam/TCR plant and a basic STIG plant, as functions of the steam/air ratio S (after Lloyd [5]).

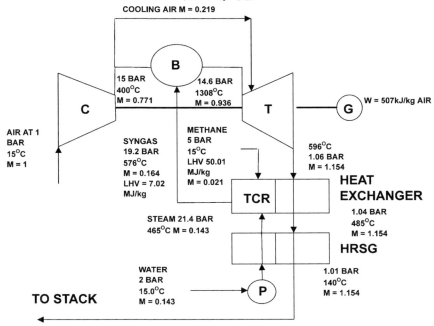

Fig. 8.11. Detailed calculation of a steam/TCR plant (after Lloyd [5]). Princeton University Library.

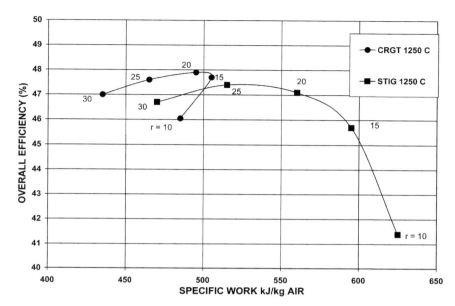

Fig. 8.12. Plots of overall efficiencies against specific work, for steam/TCR and basic STIG plants (after Lloyd [5]), Princeton University Library.

(ii) a semi-closed cycle (B3) in which part of the exhaust gas is recirculated to the reformer, together with the fuel supply, to form a new syngas fuel (the process described in Section 8.5.4).

In both cases heat is taken from the exhaust gases to 'feed' the reaction process, enhancing the 'heating value' of the resulting modified fuel, which is then fed to the combustion chamber. But the main thermodynamic feature is that the exergy loss in the final exhaust gas is thus reduced and the efficiency increased.

8.6.2.1. The steam/TCR cycle

Fig. 8.9 shows a chemically recuperated cycle [B1] of the first type, i.e. chemical recuperation with steam reforming (steam/TCR).

We first refer back to Section 6.2.2, which gave a simplified first law of analysis of a modified STIG type cycle presented by Lloyd [5]. He described an additional heat exchanger in the STIG cycle raising the enthalpy of the air entering the combustion chamber. The exhaust gas from the turbine thus first passed through this recuperator, effectively reducing the external 'heat supplied' to the combustion chamber. Lloyd argued that the heat transferred in the reformer of the steam/TCR plant performs a similar function to that of the recuperator in the modified STIG cycle. Lloyd's point is illustrated in Fig. 8.10, which shows that for a given S, the efficiency of the modified cycles is higher, the amount of steam taken out of the turbine exhaust being greater.

Lloyd's detailed computation for a steam/TCR cycle is shown in Fig. 8.11. Here the main thermodynamic parameters have been specified: pressure ratio 15, turbine entry temperature 1250°C (after turbine cooling), which with the selected turbomachinery efficiencies leads to a recuperation temperature and a pressure level of about 600°C and 15 bar, respectively. These enable the molal concentrations after reforming to be calculated, as explained in Section 8.5.2. α (the conversion rate) is determined as 0.373 and β as 0.190, so the concentrations after reforming are as follows: CH_4, 8.1%; CO, 0.4%; H_2, 19%; CO_2, 4.5%; H_2O, 68%. Thus, with 37.3% of the CH_4 converted, it follows that the heat transferred from the exhaust gas is about 110 kJ and the heating value of the resultant reformed syngas is 0.164 $[CV]_0 = 1.15$ MJ, where $[CV]_0 = 7.02$ MJ/kg is the syngas calorific value. Calculation of the remaining part of the cycle is straightforward.

The heating value of the gas supplied for combustion is enhanced by about 10% (although the calorific value is substantially reduced compared to the methane supplied, from some 50 to 7 MJ/kg). This is mainly due to the large concentration of hydrogen, as indicated in the equilibrium concentrations of the gases following the reforming. However, the thermal efficiency of the cycle is given by the work output divided by the calorific value of the original methane fuel supplied and is 47.6%.

Lloyd carried out a range of similar calculations, for differing thermodynamic parameters; the results are presented in Fig. 8.12 in comparison with those for a basic STIG cycle with the same parameters of pressure ratio and maximum temperature. There is indeed similarity between the two sets, with the TCR plant having a higher efficiency. It is noteworthy that both cycles obtain high thermal efficiency at quite low pressure ratios as one would expect for what are essentially CBTX recuperative gas turbine cycles.

Newby et al. [6] also studied a steam/TCR cycle with similar parameters and steam/air ratio. They calculated an efficiency of 48.7%, compared with 35.7% for a comparable CBT plant, 45.6% for a STIG plant and 56.8% for a CCGT plant, all for similar pressure ratios and top temperatures.

Fig. 8.13 shows Cycle B2, a development of Lloyd's simple steam/TCR cycle for CO_2 removal, as proposed by Lozza and Chiesa [7]. However, this is a CCGT plant in which the syngas produced by the steam reformer is cooled and then fed to a chemical absorption process. This enables both water and CO_2 in the syngas to be removed and a hydrogen rich syngas to be fed to the combustion chamber.

After allowing for the performance penalties arising from the CO_2 removal, Lozza and Chiesa estimated an efficiency of 46.1%, for a maximum gas turbine temperature of 1641 K and a pressure ratio of 15 (compared with the basic CCGT plant efficiency of 56.1%). They concluded that the plant cannot compete, in terms of electricity price, with a semi-closed combined cycle with CO_2 removal (Cycle A2).

8.6.2.2. *The flue gas thermo-chemically recuperated (FG/TCR) cycle*

A second type of CRGT plant involving modification of the fuel before combustion (Cycle B3) is shown in Fig. 8.14. Now some part of the exhaust from the turbine (which contains water vapour) is recirculated to the reformer where the fuel is modified. Thus this FG/TCR cycle has an element of the semi-closed cycle plus modification of the combustion process. The chemical process involved in this cycle has been described in Section 8.5.4, but there is now no simple comparison that can be made between the FG/TCR cycle and the basic STIG cycle, as described in Section 8.6.2.1.

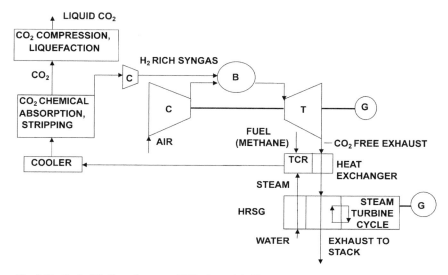

Fig. 8.13. Cycle B2. Complex steam/TCR plant with CO_2 removal (after Lozza and Chiesa [7]).

A discussion of the merits of this cycle was given by Rabovitser et al. [8] who suggested that the reforming rate of the natural gas can be increased by low oxygen content in the reacting mixture, so that the gas turbine combustor has to operate just above the stoichiometric air fuel ratio. They also suggested that for natural gas/FGR reforming the recycling coefficient (recycled stream to non-recycled stream) should be greater than unity. They quote a cycle calculation for reforming at 20 bar and 900 K with a recycling coefficient of 1.2; the reformed fuel contains only 14.2% of combustible gas (8.4% hydrogen, 2.4% CO and 3.4% CH_4). Its calorific value is only about 2.7 MJ/kg

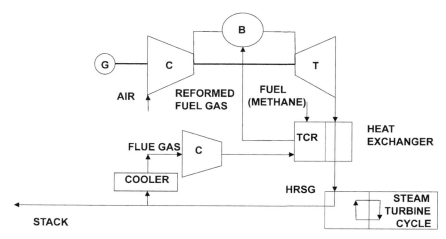

Fig. 8.14. Cycle B3. Chemically recuperated plant with flue-gas reforming (after Newby et al. [6]).

compared with 50 MJ/kg for methane itself, but of course there is now an even larger flow of combustible gas that goes to the combustor so the 'heating value' is slightly increased.

In another example Newby et al. [6] calculated a cycle with the reformer operating at comparable pressure and temperature but with a higher recycling rate of 1.7, leading to a conversion rate of $\alpha = 0.56$ (this is closer to the conversion rate of Lloyd's steam/TCR cycle, $\alpha = 0.373$, described in the last section). A thermal efficiency of 38.7% is claimed for this FG/TCR cycle, slightly greater than the simple CBT cycle efficiency of 35.7% but much less than the calculated efficiency for the steam/TCR cycle (48.7%) and a comparable STIG cycle (45.6%).

Clearly, these figures suggest that the plant is very sensitive to the amount of flue gas recycled. There appears to be no full parametric or economic calculation published in the literature for this FG/TCR cycle, which suggests that it has not been considered as an attractive option.

8.6.3. Cycles C burning non-carbon fuel (hydrogen)

Obviously, use of a non-carbon fuel—usually containing hydrogen—obviates the need for any carbon dioxide extraction and disposal. These cycles are listed in Table 8.1C, and the associated Figs. 8.15–8.17.

Fig. 8.15 shows a simple gas turbine plant (Cycle C1) supplied with a mixture of hydrogen and nitrogen for combustion in air; a cooler is shown but a bottoming steam cycle may be added (see later, C2, C3).

Jackson et al. [9] have presented calculations of thermal efficiency for this simple hydrogen fuelled CBT cycle, first with very low nitrogen content in the fuel and secondly with 50/50 hydrogen/nitrogen. For the first case they find relatively little change in

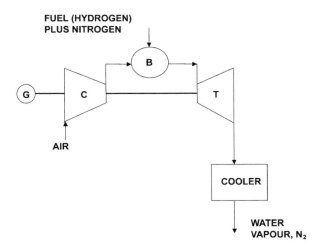

Fig. 8.15. Cycle C1. CBT plant with non-carbon fuel (hydrogen/nitrogen mixture).

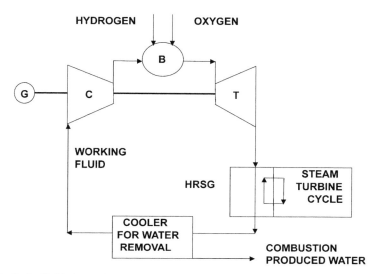

Fig. 8.16. Cycle C2. Hydrogen fuelled CCGT plant with closed upper cycle (after Bannister et al. [10]).

efficiency from that of a reference engine with the same pressure ratio and maximum temperature and burning natural gas. For the second case, with a substantial extra nitrogen flow through the turbine—giving extra turbine work—the question of whether the fuel is supplied at combustion chamber pressure becomes critical, i.e. whether the cycle has to be debited with the nitrogen compression work.

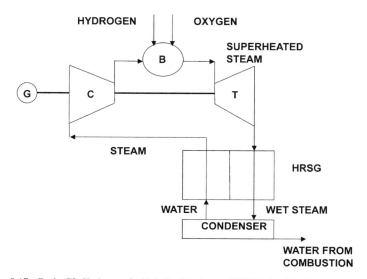

Fig. 8.17. Cycle C3. Hydrogen fuelled 'Rankine' type CCGT (after Bannister et al. [10]).

Bannister et al. [10] made a study of the hydrogen fuelled CCGT plant (Cycle C2), with a closed upper 'gas turbine' cycle (Fig. 8.16). A number of different working fluids were used in the latter, the water produced in combustion being separated and extracted downstream of the HRSG. Again there is relatively little variation of efficiency with choice of the upper cycle working fluid, each of which has some practical limitations, but that with steam as the working fluid offers highest efficiency, approaching 60% (HHV).

Bannister et al. then considered a novel 'Rankine' type hydrogen fired cycle (Cycle C3), as shown in Fig. 8.17. Low pressure wet steam leaving the turbine in the 'gas turbine' upper cycle then enters the hot side of the HRSG. After leaving the HRSG as wetter steam this mainstream flow enters the condenser. After condensation, some water, equal in mass flow to that produced in combustion (m per unit flow at entry), is then discharged. The rest (unit) flow is pumped back into the cold side of the HRSG to receive heat from the $(1 + m)$ wet steam stream. Within the HRSG, this unit water flow passes through

(a) an economiser,
(b) an evaporator to leave as saturated steam, and
(c) a superheater to impart a margin of superheat before entry to the combustion chamber.

This superheated steam then acts as a moderator for the hydrogen/oxygen combustion, which takes place at high pressure, 166 bar in the original study. Two subsequent reconfigurations of the cycle changed this high pressure to 365 and 250 bar, respectively, the same general cycle approach being followed but with some added cooling streams. The Westinghouse group concluded that a cycle efficiency of 60% (HHV) could be achieved with this Rankine type cycle.

Further detailed studies of several complex hydrogen fuelled cycles, including the 'Rankine' cycle C3, have been made by Japanese authors, e.g. Sugisita et al. [11]. Their preference is for a 'topping/extraction' cycle. In this cycle, the mainstream flow from the combustor in the upper cycle, after passing through an HP steam turbine, gets cooled in the first of the two heat exchangers, from a superheated state to the saturation condition. The flow is then split, one stream expanding further to condenser pressure, with the combustion product water flow (m) being discharged. The remainder of this stream is pumped up, recuperated by the second of the two heat exchangers, expanded again in another turbine and then mixed with the remaining topping cycle flow. Sugisita et al. claim over 60% efficiency for this so-called Jericha cycle.

8.6.4. Cycles with modification of the oxidant in combustion

We next consider a number of plants in which the combustion process is modified by changing the oxidation of the fuel, Table 8.1D and Figs. 8.18–8.20. The first group (D1, D2 and D3) are plants with PO—insufficient air is supplied to the PO reactor, less than that required to produce stoichiometric combustion. The second group (D4, D5 and D6) are plants where air is replaced as the oxidant by pure oxygen which is assumed to be available from an air separation plant.

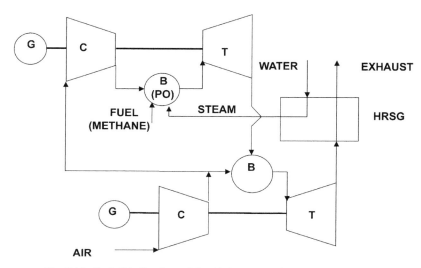

Fig. 8.18. Cycle D1. Simple partial oxidation plant (after Newby et al. [12]).

8.6.4.1. Partial oxidation cycles

A simple PO plant (D1). Fig. 8.18, after Newby et al. [12], shows a simple PO plant, of the type listed as D1 in Table 8.1D. In this plant insufficient air is supplied to the PO reactor, less than that required for producing stoichiometric combustion. After expansion in the PO turbine the fuel gas is fed to the main turbine combustor where additional air is also supplied for complete combustion.

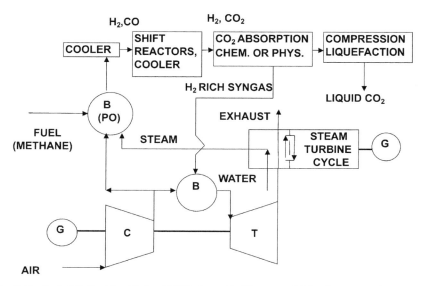

Fig. 8.19. Cycle D2. Partial oxidation CCGT plant with CO_2 removal (after Lozza and Chiesa [13]).

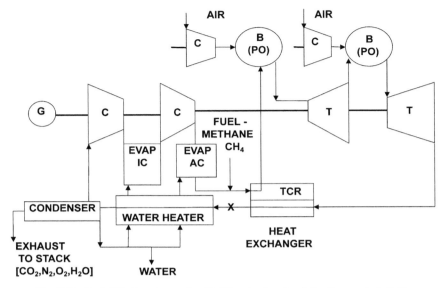

Fig. 8.20. Cycle D3. Complex cycle with PO and reforming (after Harvey et al. [14]).

A feature of this cycle is the reduction in compressor air flow for the same size of main expansion turbine. The figure shows air for the PO turbine taken from the discharge of the main compressor, but it may be taken straight from atmosphere. Note also that steam is raised for injection into the PO reactor and Newby et al. suggested that some of the steam raised in the HRSG may also be used to cool the PO turbine. The chemical reactions for the PO reactor of this case were described in Section 8.5.3.

Newby and his colleagues provided some calculations of the performance of this partial oxidation cycle. They show that a major parameter in the performance of the PO cycle is the PO turbine inlet pressure, and listed calculations for three values of this pressure: 45 bar, 60 and 100 bar. Their results for the composition of the gas streams round the plant (from the 60 bar calculation, which gave 49.3% for 335 MW) are given in Table 8.2.

Table 8.2
Newby's calculations

Stream	PO reactor outlet	PO turbine outlet	Combustion turbine outlet	Stack
Temperature (°C)	1316	773	608	98
Pressure (bar)	59.3	15.9	1.05	1.01
Mole fractions				
O_2	0	0	0.0571	0.0574
N_2	0.4646	0.3002	0.5421	0.5426
CO	0.0780	0.0504	0	0
CO_2	0.0430	0.0276	0.0444	0.0443
H_2O	0.2529	0.5173	0.3498	0.3491
H_2	0.1588	0.1006	0	0
CH_4	0	0	0	0
Mass flow kg/h	0.56×10^6	0.81×10^6	1.77×10^6	1.77×10^6

Of course, there is no methane at exit from the PO reactor, and no oxygen. The hydrogen content is quite high, over 15% and comparable to that in Lloyd's example of the steam/TCR cycle, but the CO content is also nearly 8%. It is interesting to note that the calculated equilibrium concentrations of these combustible products from the reactor are reduced through the PO turbine (because of the fall in temperature) before they are supplied to the gas turbine combustor where they are fully combusted, but it is more likely that the concentrations would be frozen near the entry values.

Newby et al. found that increasing the PO turbine pressure resulted in higher steam flow (for a given pinch point temperature difference in the HRSG), increased PO turbine power and overall plant efficiency. However, at the highest pressure of 100 bar attempts to increase the steam flow further resulted in incomplete combustion in the main combustor and the overall thermal efficiency did not increase substantially at this pressure level.

PO plant with CO_2 removal (D2). Lozza and Chiesa [13] have proposed a partial oxidation CCGT plant with carbon dioxide removal, Cycle D2 of Table 8.1D, and this is shown in Fig. 8.19. Now the syngas from a first PO reactor is cooled and fed to an additional shift reactor and then to a chemical or physical absorption plant. CO_2 can thus be removed and hydrogen rich syngas fed to the main combustion chamber of the gas turbine plant, the exhaust gases from which pass through an HRSG, producing steam for a bottoming steam cycle and the PO reactor. Lozza and Chiesa calculated a hydrogen molal fraction of nearly 50% after the shift reaction and CO_2 removal. The plant efficiency drops to 48.5% from the figure of 56.1% for a basic CCGT plant. The cost of electricity produced was estimated to be comparable to that of the semi-closed plant of Cycle A2, i.e. an increase of about 40% on that of the electricity produced by the basic CCGT plant.

Complex cycle with partial oxidation and reforming (D3). An ingenious cycle has been proposed by Harvey et al. [14] which combines both successive PO and chemical recuperation in a semi-closed cycle, as illustrated in Fig. 8.20. Recycled exhaust gases containing CO_2, H_2O and N_2 act as oxygen carriers. Partial combustion (or oxidation) takes place in successive combustors to which air is admitted (three in the proposed cycle, but only two, for illustration, in the figure). Expansion downstream of the combustors takes place through successive turbines. The exhaust gas from the last turbine is then recycled to a 'FG' reformer to which methane is admitted (the gas has been compressed and evaporatively water-cooled in three stages). However Rabovitser et al. [7], in a discussion of this cycle, argued that since the water content of the exhaust gas streams is high (5.25 mol of H_2O, 1 mol of CO_2 and 7.52 mol of N_2 per mole of CH_4 supplied) the reformer is more a steam reformer than a FG reformer.

The cycle is complex but highly efficient. This high efficiency comes from the nature of the cycle (essentially a complex version of the intercooled, reheated, recuperative CICICIBTBTBTX plant described in Chapter 3). As Harvey et al. argue, the combustion irreversibility is reduced in the successive partial combustion steps, a move towards reversible isothermal combustion.

Harvey et al. gave a parametric calculation of the thermal efficiency of this plant, as a function of turbine inlet temperature, the reformer pinch point temperature difference and the pressure level in the reformer (the compressor overall pressure ratio, *r*).

Their calculations show remarkably high overall efficiency, ranging from 56% at 1300 K to over 64% at 1500 K (with r between 20 and 25).

8.6.4.2. *Plants with combustion modification (full oxidation)*

A number of plants have been proposed in which pure oxygen is used for combustion, usually in combination with the concept of cycle semi-closure.

CBT and CCGT plants with full oxidation (D4, D5). We next consider two semi-closed cycles for CO_2 removal (Cycles D4 and D5) with air replaced as the oxidant for the fuel, by pure oxygen supplied from an additional plant.

In cycle D4 [15], since the fuel is burnt with pure oxygen, the exhaust gases contain CO_2 and H_2O almost exclusively (Fig. 8.21). Cooling the exhaust below the dew point enables the water to condense and the resulting CO_2 stream is obtained without the need for chemical absorption. The expensive auxiliary plant involved in direct removal of the CO_2 is not needed, but of course there is now the additional expense of an air separation plant to provide the pure oxygen for combustion.

Cycle D5 is another variation of a CCGT plant with full oxygenation of the fuel as shown in Fig. 8.22; again it is a semi-closed cycle using pure oxygen. But now the CO_2 is abstracted after compression, which may require the use of physical absorption plant.

For cycle D4 it may be expected that the thermal efficiency will be close to that of the open CBT plant with the same pressure ratio and top temperature. For cycle D5 there will be a penalty on efficiency imposed from the extra compression of CO_2 before extraction.

The Matiant cycle (D6). Fig. 8.23 shows a more complex and ingenious version of the semi-closed cycle burning fuel with oxygen—the so-called Matiant plant [16]. A stage

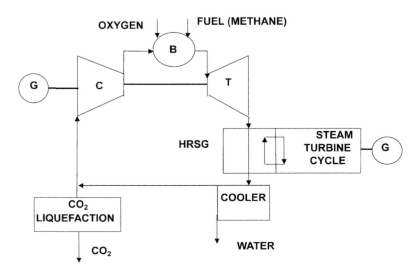

Fig. 8.21. Cycle D4. Simple CCGT plant burning methane with oxygen, and with low pressure CO_2 removal.

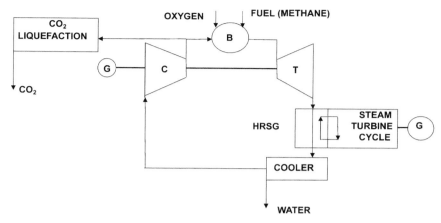

Fig. 8.22. Cycle D5. Simple CCGT plant burning methane with oxygen, and with high pressure CO_2 removal.

of reheat and three stages of compression are involved together with a recuperator. Carbon dioxide and water vapour are the working gases but both the CO_2 and H_2O formed in combustion are removed, the former through a complex compression and liquefaction process. The multiple reheating and intercooling implies that such a cycle should attain high efficiency, with 'heat supplied' near the top temperature and 'heat rejected' near the bottom temperature, coupled with CO_2 removal.

Manfrida [4] calculated a thermal efficiency of 55% for this cycle at a maximum cycle pressure of 250 bar and a combustion temperature of 1400°C.

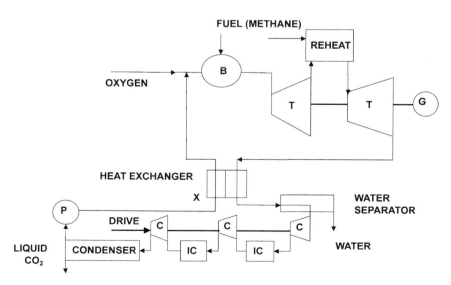

Fig. 8.23. Cycle D6. Matiant closed CICICBTBTBTX cycle burning methane with oxygen, and with CO_2 removal (after Manfrida [4]).

8.7. IGCC cycles with CO_2 removal (Cycles E)

The IGCC cycle was described in Section 7.4.2. Obviously, there is an attraction in burning cheap coal instead of expensive gas, but the IGCC plant will discharge as much carbon dioxide as a normal coal burning plant unless major modifications are made to remove the CO_2 (Table 8.1E).

As for the conventional methane burning cycles the IGCC plants can be modified
(a) for addition of CO_2 absorption equipment in a semi-closed cycle (Cycle E1);
(b) for combustion with fuel modification with extra water shift reaction downstream of the syngas production plant (Cycle E2); and
(c) for combustion with full oxidation of the syngas (Cycle E3).

Fig. 8.24 shows an example of a semi-closed plant (Cycle E1) as studied by Chiesa and Lozza [17]. The CO_2 absorption takes place downstream of the HRSG after further cooling with water removal.

Fig. 8.25 shows an example of the second open type of IGCC plant proposed (Cycle E2) with an additional shift reactor downstream of the gasifier and syngas cooling and cleansing plant. Absorption of the CO_2 is at high pressure which may require physical absorption equipment of the type described in Section 9.2.2 [3]. However, Manfrida [4] argued that it is still possible to use chemical absorption at moderately high pressure in this IGCC plant.

Finally, Fig. 8.26 shows Cycle E3—a semi-closed IGCC plant with oxygen fed to the main syngas combustion process in a semi-closed cycle [18]. Now the exhaust from the HRSG is cooled before removal of the CO_2 at low pressure, without need of complex equipment.

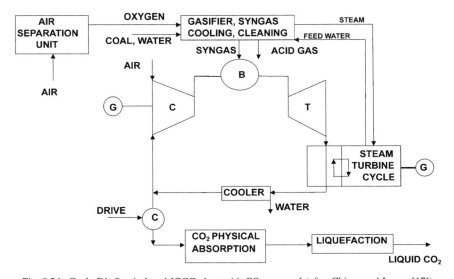

Fig. 8.24. Cycle E1. Semi-closed IGCC plant with CO_2 removal (after Chiesa and Lozza [17]).

Table 8.1E
Cycles E with modifications of IGCC plants using syngas

Description	Type	Special features	Fuel/oxidant	CO$_2$ removal	Comment
E1 Semi-closed IGCC/CO$_2$ removal	SC/IGCC	–	Syngas/air	LP physical absorption	Expensive
E2 (i) IGCC/shift/CO$_2$ removal	Open/IGCC	Extra water shift	Syngas/air	HP physical absorption	Radiation or quench cooling
E2 (ii) IGCC/shift/CO$_2$ removal	Open/IGCC	Extra water shift	Syngas/air	HP chemical absorption	Quench cooling
E3 Oxygen blown IGCC	SC/IGCC	Extra oxygen plant	Syngas/oxygen	LP extraction plus compression/liquefaction	Large oxygen consumption

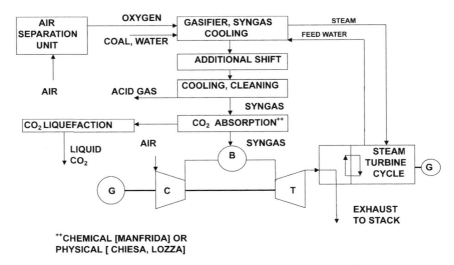

Fig. 8.25. Cycle E2. Open IGCC plant with shift reactor and CO_2 removal (after Chiesa and Consonni [3]).

8.8. Summary

The performance of these novel plants may be assessed in relation to two objectives—the attainment of good performance (high thermal efficiency and low cost of electricity produced) and the effectiveness of CO_2 removal, although the two may be coupled if a CO_2 tax is introduced.

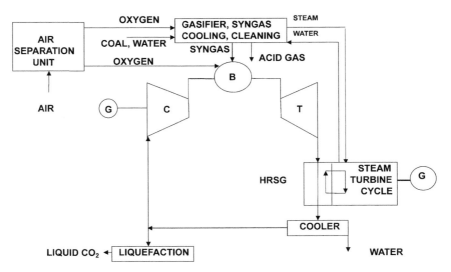

Fig. 8.26. Cycle E3. Semi-closed IGCC plant with oxygen feed and CO_2 removal (after Chiesa and Lozza [18]).

Few of these novel cycles can be compared with good modern CCGT plants operating at high turbine entry temperatures, with very high overall efficiencies approaching 60%. Some of the new cycles requiring modification of the basic CBT plant (TCR or PO) cannot match the high efficiency of the CCGTs; those that can match the overall efficiency usually involve additional processes and equipment and therefore incur an increased capital cost.

In particular, the cycles involving fuel or oxidant modification do not look sufficiently attractive for their development to be undertaken, with the possible exception of the multiple PO combustion plant proposed by Harvey et al. [14]. The Matiant plant has the advantage of relatively simple CO_2 removal and high efficiency and may prove to be attractive, but it again looks complex and expensive.

Modifications of the existing plants to sequestrate and dispose of the CO_2 will lead to a reduction in net thermal efficiency and an increase in capital cost; both these features will lead to increased cost of electricity generation. Whether these plants will be economic in comparison with conventional plants of higher efficiency and less capital cost will be determined by how much the conventional plants will have to pay in terms of a carbon tax.

Chiesa and Consonni [1,3] have made detailed studies of how a CO_2 tax would affect the economic viability of several of these cycles when a tax and CO_2 removal are introduced. Fig. 8.27 shows their results on the cost of electricity for natural gas-fired plants plotted against the level of a carbon tax (in c/kg CO_2 produced), for two of the novel cycles studied here, in comparison with an existing CCGT plant with natural gas firing.

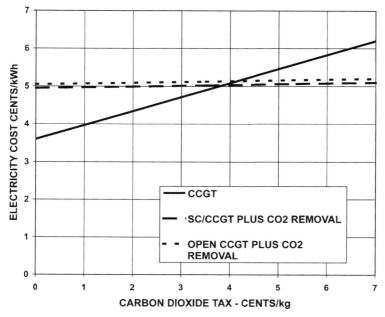

Fig. 8.27. Electricity price variation with carbon tax for (i) CCGT plant, (ii) semi-closed CCGT plant with CO_2 removal, (iii) open CCGT plant with CO_2 removal (after Chiesa and Consonni [1]).

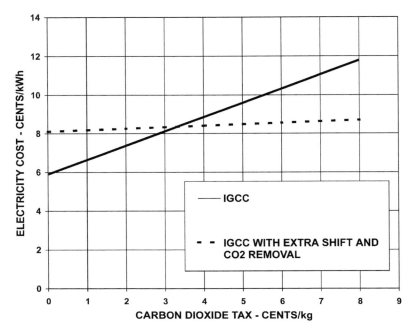

Fig. 8.28. Electricity price variation with carbon tax for (i) IGCC plant and (ii) IGCC plant with extra shift and CO_2 removal (after Chiesa and Consonni [3]).

The novel cycles are:

(i)   a natural gas-fired open CCGT plant with 'end of pipe' CO_2 removal at low pressure (Cycle A1); and

(ii) a natural gas-fired semi-closed CCGT plant with CO_2 removal by chemical absorption at low pressure (Cycle A2).

Fig. 8.28 shows a similar plot for coal fired IGCC plant with and without CO_2 removal (by extra shift reaction and CO_2 removal at high pressure (Cycle F1)).

Clearly, the carbon dioxide tax will be a dominant factor in future economic analyses of novel cycles. It would appear that a tax of about 3 c/kg of CO_2 produced would make some of the CO_2 removal cycles economic when compared to the standard basic cycles.

References

[1] Chiesa, P. and Consonni, S. (2000), Natural gas fired combined cycles with low CO_2 emissions, ASME J. Engng Gas Turbines Power 122(3), 429–436.

[2] Corti, G. and Manfrida G. (1998), Analysis of a semi-closed gas turbine/combined cycle (SCGT/CC) with CO_2 removal by amines absorption, International Conference On Greenhouse Gas Control Technologies, Interlaken.

[3] Chiesa, P. and Consonni, S. (1999), Shift reaction and physical absorption for low emission IGCCs, ASME J. Engng Gas Turbines Power 121(2), 295–305.

[4] Manfrida, G. (1999), Opportunities for high-efficiency electricity generation inclusive of CO_2 capture, Int. J. Appl. Thermodyn. 2(4), 165–175.

[5] Lloyd, A. (1991), Thermodynamics of chemically recuperated gas turbines, CEES Report 256, Centre For Energy and Environmental Studies, University Archives Department of Rare Books and Special Collections, Princeton University Library.

[6] Newby, R.A., Yang, W.C. and Bannister, R.L. (1997), Use of thermochemical recuperation in combustion turbine power systems, ASME Paper 97-GT-44.

[7] Lozza, G. and Chiesa, P. (2001), Natural gas decarbonisation to reduce CO_2 emission from combined cycle—Part II: steam–methane reforming, ASME J. Engng Gas Turbines Power 124(1), 89–95.

[8] Rabovitser, J.K., Khinkis, M.J., Bannister, R.L. and Miao, F.Q. (1996), Evaluation of thermochemical recuperation and partial oxidation concepts for natural gas-fired advanced turbine systems, ASME paper 96-GT-290.

[9] Jackson, A.J.B., Audus, H. and Singh, R. (2000), Gas turbine requirement for power generation cycles having CO_2 sequestration, ISABE-2001–1176.

[10] Bannister, R.L., Huber, D.J., Newby, R.A. and Paffenburger J.A. (2000), Hydrogen-fuelled combustion turbine cycle, ASME paper 96-GT-246.

[11] Sugisita, H., Mori, H. and Uematsu, K. (1996), A study of advanced hydrogen/oxygen combustion turbines, Unpublished MHI report.

[12] Newby, R.A., Yang, W.C. and Bannister, R.L. (1997), An evaluation of a partial oxidation concept for combustion turbine power systems, ASME Paper 97-A4-24.

[13] Lozza, G. and Chiesa, P. (2002), Natural gas decarbonisation to reduce CO_2 emission from combined cycle—Part I: Partial oxidation, ASME J. Engng Gas Turbines Power 124(1), 82–88.

[14] Harvey, S.P., Knoche, K.E. and Richter, H.J. (1995), Reduction of combustion irreversibility in a gas turbine power plant through off-gas recycling, ASME J. Engng Gas Turbines Power 117(1), 24–30.

[15] Ulizar, I. and Pilidis, P. (1996), A semi-closed cycle gas turbine with carbon dioxide–argon as working fluid, ASME paper 96-GT-345.

[16] Mathieu, P. and Nihart, R. (1999), Zero-emission MATIANT cycle, ASME J. Engng Gas Turbines Power 121(1), 116–120.

[17] Chiesa, P. and Lozza, G. (1999), CO_2 emission abatement in IGCC power plants by semi-closed cycles—Part B—with air blown combustion and CO_2 physical absorption, ASME J. Engng Gas Turbines Power 121(4), 642–648.

[18] Chiesa, P. and Lozza, G. (1999), CO_2 emission abatement in IGCC power plants by semi-closed cycles—Part A with oxygen-blown combustion, ASME J. Engng Gas Turbines Power 121(4), 635–641.

Chapter 9

THE GAS TURBINE AS A COGENERATION (COMBINED HEAT AND POWER) PLANT

9.1. Introduction

The thermodynamics of thermal power plants has long been a classical area of study for engineers. A conventional power plant receiving fuel energy (F), producing work (W) and rejecting 'non-useful' heat (Q_A) to a sink at low temperature was illustrated earlier in Fig. 1.1. The designer attempts to minimise the fuel input for a given work output because this will clearly give economic benefit in the operation of the plant, minimising fuel costs against the sales of electricity to meet the power demand.

The objectives of the designer of a combined heat and power plant are wider, for both heat and work production. Fig. 9.1 shows a CHP or cogeneration (CG) plant receiving fuel energy (F_{CG}) and producing work (W_{CG}). But useful heat ($Q_U)_{CG}$, as well as non-useful heat ($Q_{NU})_{CG}$ is now produced. Both the work and the useful heat can be sold, so the CHP designer is not solely interested in high thermal efficiency, although the work output commands a higher sale price than the useful heat output. Clearly, both thermodynamics and economics will be of importance and these are developed in Ref. [1]. A much briefer discussion of CHP is given here.

Fig. 9.2 shows how a simple open circuit gas turbine can be used as a cogeneration plant: (a) with a waste heat recuperator (WHR) and (b) with a waste heat boiler (WHB). Since the products from combustion have excess air, supplementary fuel may be burnt downstream of the turbine in the second case. In these illustrations, the overall efficiency of the gas turbine is taken to be quite low ($(\eta_O)_{CG} = W_{CG}/F_{CG} = 0.25$), where the subscript CG indicates that the gas turbine is used as a recuperative cogeneration plant.

In Fig. 9.2a, the work output from the unfired plant is shown to be equal to unity and the heat supply $F_{CG} = 4.0$. Further, it is assumed that the useful heat supplied is $(Q_U)_{CG} = 2.25$ and the unused non-useful heat is $(Q_{NU})_{CG} = 0.75$. An important parameter of this CHP plant is the ratio of useful heat supplied to the work output, $\lambda_{CG} = (Q_U)_{CG}/W_{CG} = 2.25$.

For a plant with a fired heat boiler, as in Fig. 9.2b, both the work output W_{CG} and the main heat supply $F_{CG} = F_1$ are assumed to be unaltered at 1.0 and 4.0, respectively, but supplementary fuel energy is supplied to the WHB, $F_2 = 1.5F_1 = 6.0$. The useful heat supplied is then assumed to increase to 7.2 and the non-useful heat rejected to be 1.8. Thus the parameter λ changes to 7.2.

For a site with a fixed power demand throughout the year, the unfired plant illustrated in Fig. 9.2a is suitable for summer operation when the heat load is light.

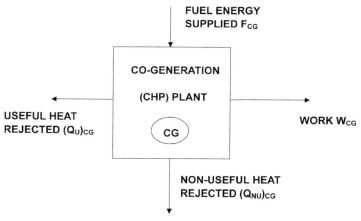

Fig. 9.1. Cogeneration (CHP) plant.

It could then be supplementarily fired in the winter when the heat load is heavier, as in Fig. 9.2b.

9.2. Performance criteria for CHP plants

9.2.1. Energy utilisation factor

For an open circuit power plant, an (arbitrary) overall efficiency has been defined as

$$\eta_O = \frac{W}{F}. \tag{9.1}$$

This criteria of performance has less relevance to a combined heat and power plant which provides heat and generates electrical power. For an open circuit gas turbine plant, a more logical criterion is the energy utilisation factor (EUF) which can be calculated as

$$EUF = \frac{W_{CG} + (Q_U)_{CG}}{F_{CG}} = (\eta_O)_{CG}(1 + \lambda_{CG}), \tag{9.2}$$

where $(Q_U)_{CG}$ is the useful heat rejected to meet the required heat load, at a temperature T_U higher than T_0, the temperature of the environment. It is preferable not to use the term efficiency for EUF, to avoid confusion with the thermal or overall efficiency.

For the unfired example, Fig. 9.2a, the efficiency is 0.25, and EUF $= (W + Q_U)/F = (1 + 2.25)/4 = 0.8125$. For the supplementary fired example of Fig. 9.2b, the efficiency remains at 0.25 but the EUF becomes

$$EUF = \frac{W + Q_U}{F_1 + F_2} = \frac{1 + 7.2}{4 + 6} = 0.82.$$

It must be remembered that work is difficult to produce and highly priced, whereas the useful heat is a lower grade, lower priced product from the plant. The energy utilisation

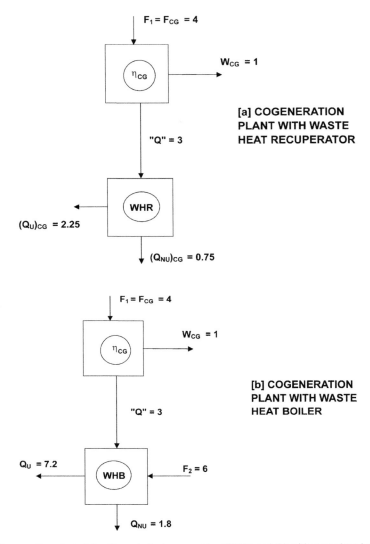

Fig. 9.2. Cogeneration plants (a) with waste heat recuperator (WHR) and (b) with waste heat boiler (WHB).

factor is thus not entirely satisfactory as a criterion of performance as it gives equal weight to W and Q_U. A 'value-weighted' EUF is therefore sometimes used, accounting for the different pricing of electrical power and heat load. If the sale price of electrical power is Y_E (£/kWh), that of the heat load is Y_H (£/kWh) and the price of fuel is ξ (£/kWh) then the 'value-weighted' EUF can be calculated as

$$(EUF)_{VW} = \frac{Y_E W + Y_H Q_U}{\xi F} = \frac{Y_E (\eta_O)_{CG}}{\xi}\left[1 + \frac{Y_H}{Y_E}\lambda\right]. \qquad (9.3)$$

9.2.2. Artificial thermal efficiency

A second criterion of performance sometimes used is an 'artificial' thermal efficiency (η_A) in which the energy in the fuel supply to the CHP plant is supposed to be reduced by that which would be required to produce the heat load (Q_U) in a separate 'heat only' boiler of efficiency (η_B), i.e. by (Q_U/η_B). The artificial efficiency (η_A) is then given by

$$\eta_A = \frac{W}{F - (Q_U/\eta_B)} = \frac{(\eta_O)_{CG}}{1 - \left(\dfrac{Q_U}{\eta_B F}\right)}, \tag{9.4}$$

where $(\eta_O)_{CG}$ is the overall efficiency of the CHP plant.
For the unfired plant of Fig. 9.2a and taking $\eta_B = 0.90$, the artificial efficiency would be

$$\eta_A = \frac{0.25}{1 - \dfrac{2.25}{(0.9)4}} = \frac{0.25}{0.375} = 0.666.$$

For the supplementary fired plant of Fig. 9.2b, the artificial efficiency would be

$$\eta_A = \frac{W}{(F_1 + F_2) - Q_U/\eta_B} = \frac{1}{10 - \dfrac{7.2}{0.9}} = 0.5.$$

9.2.3. Fuel energy saving ratio

A third performance criterion developed for combined heat and power plant involves comparison between the fuel required to meet the given loads of electricity and heat in the CHP plant with that required in a 'reference system'. The latter involves conventional plants that meet the same load demands (indicated by subscript D), for example, in a conventional electric power station and in a 'heat only' boiler.

Such a 'reference system' is shown in Fig. 9.3a. The overall efficiency of the conventional electric power plant is η_C (for simplicity the subscript O for overall efficiency is dropped from here onwards); the (demand) electrical load is unity. The ratio of heat to electrical demands is λ_D, so that the demand heat load is taken as λ_D. The efficiency of the 'heat only' boiler is η_B so the fuel energy required for the boiler is (λ_D/η_B), i.e. there are heat losses $\lambda_D[(1/\eta_B) - 1]$ involved before heat is delivered to district or process heating.

A CHP system meeting the same power and heat demands (1, λ_D) is shown in Fig. 9.3b; it is implied that this cogeneration plant is perfectly matched, delivering the required (1, λ_D) precisely, using a WHR.

The total fuel energy required in the reference system is

$$F_{REF} = \frac{1}{\eta_C} + \frac{\lambda_D}{\eta_B}, \tag{9.5}$$

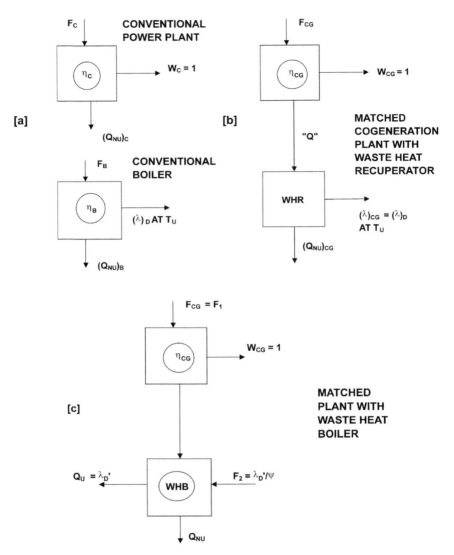

Fig. 9.3. (a) Reference system, (b) matched CHP plant with WHR, and (c) matched CHP plant with WHB.

and the fuel saving in using a CHP plant is therefore

$$\Delta F = F_{\text{REF}} - F_{\text{CG}} = \frac{1}{\eta_{\text{C}}} + \frac{\lambda_{\text{D}}}{\eta_{\text{B}}} - \frac{1}{\eta_{\text{CG}}}. \tag{9.6}$$

The fuel energy savings ratio (FESR) is then defined as the ratio of the saving (ΔF) to the fuel energy required in the conventional plants,

$$\text{FESR} = \frac{\Delta F}{F_{\text{REF}}} = \left(\frac{1}{\eta_{\text{C}}} + \frac{\lambda_{\text{D}}}{\eta_{\text{B}}} - \frac{1}{\eta_{\text{CG}}} \right) \bigg/ \left(\frac{1}{\eta_{\text{C}}} + \frac{\lambda_{\text{D}}}{\eta_{\text{B}}} \right) = 1 - \frac{(\eta_{\text{C}}/\eta_{\text{CG}})}{1 + (\lambda_{\text{D}} \eta_{\text{C}}/\eta_{\text{B}})}. \tag{9.7}$$

The above simple analysis has to be modified for a supplementary fired CHP plant such as that shown in Fig. 9.3c, meeting a unit electrical demand and an increased heat load λ'_D. The 'reference system' fuel energy supplied is now

$$F'_{REF} = \frac{1}{\eta_C} + \frac{\lambda'_D}{\eta_B}. \tag{9.8}$$

The CHP plant now requires a fuel energy supply of

$$F' = F_1 + F_2 = (1/\eta_{CG}) + F_2, \tag{9.9}$$

where $F_2 = \lambda'_D/\psi$ is the supplementary fuel energy supplied to the WHB, so that

$$EUF = \frac{(1 + \lambda'_D)}{\left(\dfrac{1}{\eta_{CG}} + \dfrac{\lambda'_D}{\psi}\right)}. \tag{9.10}$$

The quantity ψ requires discussion. The steady flow energy equation for the WHB is

$$M_{f2}h_{f0} + H_{P4} = \lambda'_D + H_{P'S}, \tag{9.11}$$

where 4 and S are the entry and exit states, P refers to products entering (i.e. at exit from the turbine), P' refers to products after the supplementary combustion and $M_{f2}h_{f0}$ is the enthalpy flux of the entering fuel. For a corresponding calorific value experiment at temperature T_0, again with products P entering and products P' leaving,

$$M_{f2}h_{f0} + H_{P0} = M_{f2}[CV]_0 + H_{P'0}. \tag{9.12}$$

From these two equations, eliminating $M_{f2}h_{f0}$,

$$\lambda'_D = M_{f2}[CV]_0 + (H_{P4} - H_{P0}) - (H_{P'S} - H_{P'0}), \tag{9.13}$$

so that

$$\psi = \frac{\lambda_{D'}}{F_2} = 1 + \frac{\{(H_{P4} - H_{P0}) - (H_{PS'} - H_{P'0})\}}{F_2}. \tag{9.14}$$

where $(H_{PS'} - H_{P'0})$ is the new 'heat loss' in the stack $(Q)'_{NU}$, and this will usually be less than $(H_{P4} - H_{P0})$, so that ψ will be greater than unity (it is not a boiler efficiency). We shall not determine ψ here but give it parametric values of 1.2 and 1.5 in the later calculations.

The fuel savings for the supplementary fired plant are given by

$$\Delta F' = \left(\frac{1}{\eta_C} + \frac{\lambda_{D'}}{\eta_B}\right) - \left(\frac{1}{\eta_{CG}} + \frac{\lambda_{D'}}{\psi}\right), \tag{9.15}$$

and the fuel savings ratio is

$$FESR' = \frac{\Delta F'}{F_{REF}} = 1 - \left\{\left(\frac{1}{\eta_{CG}} + \frac{\lambda_{D'}}{\psi}\right) \Big/ \left(\frac{1}{\eta_C} + \frac{\lambda_{D'}}{\eta_B}\right)\right\}. \tag{9.16}$$

By way of numerical illustration of the fuel savings ratio, we consider the two plants illustrated in Fig. 9.2. For the unfired plant of Fig. 9.2a, taking $\eta_C = 0.4$ and $\eta_B = 0.9$ and with $\lambda_D = \lambda_{CG} = 2.25$,

$$\text{FESR} = 1 - \frac{(0.4/0.25)}{1 + (2.25 \times 0.4/0.9)} = 0.2.$$

For the supplementary fired plant of Fig. 9.2b with $\lambda_{D'} = 7.2$ and with the parameter ψ taken as 1.2, so that $F_2 = 6$, the fuel energy savings ratio is

$$\text{FESR}' = 1 - (4 + 6)/(2.5 + 7.2/0.9) = 0.048.$$

Thus the FESR is less attractive when there is a large heat load and a WHB with supplementary firing is used. In general, the FESR is probably the most useful of the CHP plant performance criteria as it can be used directly in the economic assessment of the plant [1].

9.3. The unmatched gas turbine CHP plant

In general, a gas turbine CHP plant may not exactly match the electricity and heat demands. A plant with a recuperator may meet the heat load ($(Q_U)_{CG} = \lambda_D$) but not the power load ($W_{CG} < W_D = 1$) so extra power from the grid is required (W_C) as illustrated in Fig. 9.4. Following a procedure similar to that given in Section 9.2.3 it may be shown [1] that the performance parameters for the total plant are then

$$\text{EUF} = \frac{(1 + \lambda_D)}{F'} = \frac{(1 + \lambda_D)(1 - \eta_{CG})\eta_C}{(1 - \eta_{CG}) + \left[1 + \left(\frac{(Q_{NU})_{CG}}{\lambda_D}\right)\right](\eta_C - \eta_{CG})\lambda_D}, \qquad (9.17)$$

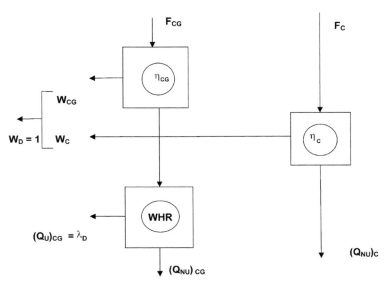

Fig. 9.4. Unmatched CHP plant taking power from the grid.

$$\text{FESR}' = 1 - \left(\frac{F'}{F_{\text{REF}}}\right)$$

$$= 1 - \eta_{\text{B}}\left\{\frac{(1 - \eta_{\text{CG}}) + \left[1 + \left(\frac{(Q_{\text{NU}})_{\text{CG}}}{\lambda_{\text{D}}}\right)\right](\eta_{\text{C}} - \eta_{\text{CG}})\lambda_{\text{D}}}{(1 - \eta_{\text{CG}})(\eta_{\text{B}} + \eta_{\text{C}}\lambda_{\text{D}})}\right\}. \qquad (9.18)$$

$(Q_{\text{NU}})_{\text{CG}}$ is usually limited by the allowable stack temperature T_{S}. As a fraction of the heat supplied to the cogeneration plant it remains constant in this application.

For an unmatched gas turbine CHP plant, meeting a power load ($W_{\text{CG}} = W_{\text{D}} = 1$) but not the heat load $(Q_{\text{U}})_{\text{CG}} < \lambda_{\text{D}}$, increased useful heat may be obtained by firing the WHB, as explained in Section 9.2.3, and illustrated in Fig. 9.3c.

9.4. Range of operation for a gas turbine CHP plant

We now illustrate numerically the full range of operation of a gas turbine CHP plant,
(i)   with a recuperator (unfired) and
(ii) with a WHB (fired).

A gas turbine plant with an overall efficiency $\eta_{\text{CG}} = 0.25$ matching a heat load $\lambda_{\text{CG}} = 2.25$ is again considered as the 'basic' CHP plant; also implied is a non-useful heat rejection ratio $(Q_{\text{NU}})_{\text{CG}}/F_{\text{CG}} = [1 - (\eta_{\text{CG}})(\lambda_{\text{CG}} + 1)] = 3/16$. For FESR calculations, we again take the conventional plant efficiency as 0.4 and the conventional boiler efficiency as 0.9. At the fully matched condition these assumptions previously led to EUF = 0.8125 and FESR = 0.2.

We next calculate EUF and FESR over a range of heat to power ratios $\lambda_{\text{D}} \neq \lambda_{\text{CG}}$.
(i)   For the plant with a WHR only, for $\lambda_{\text{D}} < \lambda_{\text{CG}}$, the power is taken via the grid from a conventional power plant. Thus Eqs. (9.17) and (9.18) yield

$$(\text{EUF}) = \frac{0.3(1 + \lambda_{\text{D}})}{0.75 + 0.2\lambda_{\text{D}}}, \qquad (9.19)$$

$$(\text{FESR}) = \frac{0.12\lambda_{\text{D}}}{0.675 + 0.3\lambda_{\text{D}}}. \qquad (9.20)$$

EUF and FESR are plotted against λ_{D} on the left hand side of Fig. 9.5. $(Q_{\text{NU}})_{\text{CG}}/F_{\text{CG}}$ is constant at 3/16 over the range from $\lambda_{\text{D}} = 0$ to 2.25, since the operation of the CG plant remains the same.
(ii) For the plant with a WHB, and for the demand λ'_{D} exceeding 2.25, Eqs. (9.10) and (9.16) give the values of EUF' and FESR' as follows:
      for $\psi = 1.2$,

$$\text{EUF}' = \frac{1.2(1 + \lambda'_{\text{D}})}{4.8 + \lambda'_{\text{D}}}, \qquad (9.21)$$

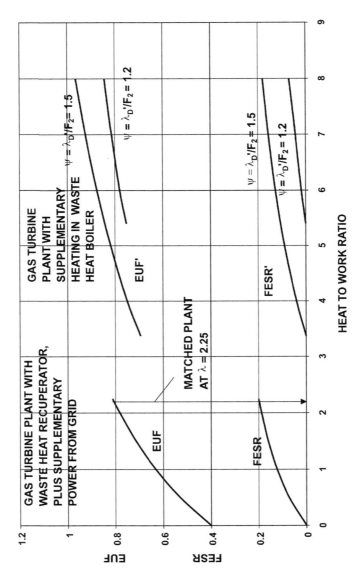

Fig. 9.5. Performance of unmatched CHP plants, with WHR and with WHB, for varying heat to work ratio (after Ref. [1]).

$$\text{FESR}' = \frac{(0.1\lambda'_D - 0.54)}{(0.9 + 0.4\lambda'_D)};$$

(9.22)

and for $\psi = 1.5$,

$$\text{EUF}' = \frac{1.5(1 + \lambda'_D)}{6 + \lambda'_D},$$

(9.23)

$$\text{FESR}' = \frac{(0.24\lambda'_D - 0.81)}{(1.35 + 0.6\lambda'_D)}.$$

(9.24)

The values of EUF' and FESR' under the conditions of λ'_D greater than 2.25 are plotted on the right hand side of Fig. 9.5. Now fuel savings appear for $\lambda'_D > 5.4$ with $\psi = \lambda'_D/F_2 = 1.2$, and for $\lambda'_D > 3.375$ with $\psi = \lambda'_D/F_2 = 1.5$.

But the amount of unused heat now varies with λ_D and is given by

$$(Q_{NU})_{CG}/F_{CG} = (3/4) + \lambda'_D[(1/\psi) - 1]/4,$$

$$(Q_{NU})_{CG}/F_{CG} = (18 - \lambda'_D)/24 \qquad \text{for } \psi = 1.2,$$

$$(Q_{NU})_{CG}/F_{CG} = (9 - \lambda'_D)/12, \qquad \text{for } \psi = 1.5.$$

These are plotted against λ_D in Fig. 9.6. Clearly, the calculations lose validity when $\lambda'_D = 18$ for $\psi = 1.2$, and when $\lambda'_D = 9$ for $\psi = 1.5$. However, if the exhaust stack

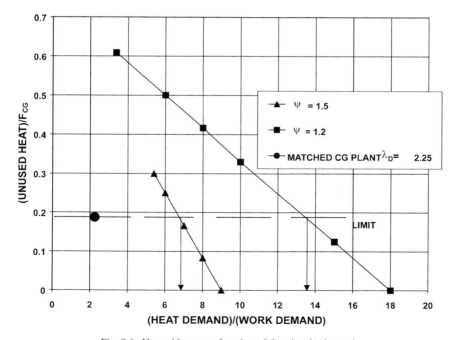

Fig. 9.6. Unused heat as a function of (heat/work) demand.

temperature is limited, at the level corresponding to $(Q_{NU})_{CG}/F_{CG} = 3/16$ as in the basic plant, then corresponding limits on λ'_D are 27/4 for $\psi = 1.5$ and 27/2 for $\psi = 1.2$.

9.5. Design of gas turbines as cogeneration (CHP) plants

Both the heat to work ratio λ_{CG} and the various performance parameters such as EUF and FESR are affected by the choice of design parameters within a gas turbine. However, for the gas turbine with a WHR, the range of λ_{CG} that can be achieved by varying these parameters is not large and operation may have to involve firing a WHB, or running in parallel with conventional plants, as explained earlier. But some variation in λ_{CG} can be achieved by varying the 'internal' design parameters (e.g. pressure ratio and turbine inlet temperature), achieving matched operation for each of the different designs, i.e. by varying λ_{CG} to match λ_D. Porter and Mastanaiah [2] calculated λ_{CG} for a gas turbine with a WHR supplying process steam at p_P, T_P. Plots of the heat to work ratio λ_{CG} against T_P are shown in Fig. 9.7, for a maximum temperature of 1200 K and various pressure ratios, and with a limit on the stack temperature and compressor and turbine efficiencies of 0.9.

The EUF and FESR are then simple to derive and typical area plots of the range of EUF and FESR against the derived λ_{CG}, for gas turbines with varying practical design parameters, are illustrated in Fig. 9.8.

It is concluded that such simple gas turbines with WHRs have good energy utilisation at $\lambda_{CG} \approx 1$ with respectable FESR. The introduction of a WHB will move the operable area to higher values of λ, usually with comparable EUFs but lower FESRs, as has been illustrated in the examples calculated in Section 9.2.

9.6. Some practical gas turbine cogeneration plants

There are many gas turbine CHP plants in operation for a range of purposes and applications. Here we describe the salient features of two such plants, each operating with a WHR but also with supplementary firing which can be introduced to meet increased heat demands.

9.6.1. The Beilen CHP plant

A gas turbine CHP scheme, with a heat recovery steam generator producing process steam, operates at the DOMO plant at Beilen in the Netherlands. The plant, which produces dairy products, originally took its electric power (up to 3.2 MW) from the grid and its heat load was met by two gas-fired boilers with a steam production of 25 t/h at 13 bar.

The CHP plant which replaced these two separate energy supplies is based on a Ruston TB gas turbine (rated at 3.65 MW) which can meet the electrical demand of 3.2 MW and is connected to the grid so that excess electrical power can be sold. By providing full gas power, up to 12 t/h of saturated steam can be produced at 191°C and 13 bar. Five supplementary gas burners can be engaged to increase the steam

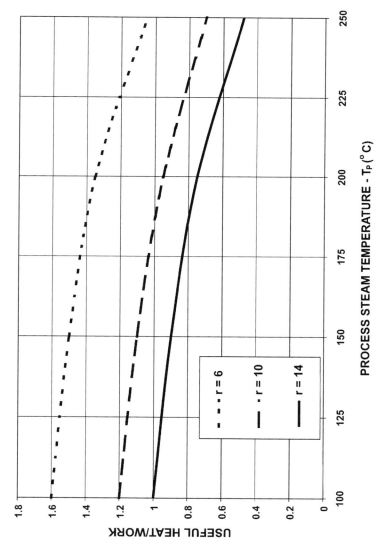

Fig. 9.7. (Useful heat)/work as a function of process steam temperature (after Porter and Mastanaiah [2]).

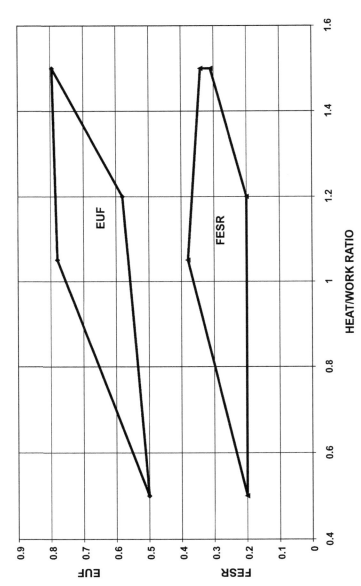

Fig. 9.8. Range of EUF and FESR for a matched CHP gas turbine plant (after Porter and Mastanaiah [2]).

production to 35 t/h. Gases leave the exhaust stack at 138°C under maximum load conditions.

For the first operating condition (HRSG unfired) the heat load is estimated at 7.5 MW. For the second condition (HRSG fired) when 35 t/h of saturated steam is raised, the heat load is 23 MW. The values of heat to work ratios (λ_D) are thus

$$\left(\frac{7.5}{3.2}\right) = 2.34, \text{ and } \left(\frac{23}{3.2}\right) = 7.19, \text{ respectively.}$$

Other parameters for the plant operating conditions—of HRSG unfired (WHR) and HRSG fired (WHB)—are as follows:

Alternator power output	3.2 MW
Airmass flow rate	20.45 kg/s
Pressure ratio	7:1
Maximum temperature	890°C
Thermal efficiency	0.23

Heat recovery steam generator

Unfired	Steam (saturated) mass flow rate	12 t/h
	Steam pressure	13 bar
Fired	Steam (saturated) mass flow rate	35 t/h
	Steam pressure	13 bar

	WHR	WHB ($\psi = 1.34$)
λ	2.34	7.19
EUF	0.77	0.85
FESR	0.147	0.075 ($\eta_C = 0.4$, $\eta_B = 0.9$)

A full description of this plant is given in Ref. [1].

9.6.2. The Liverpool University CHP plant

A gas turbine CHP scheme which operates at Liverpool University, UK, consists of a Centrax 4 MW (nominal) gas turbine with an overall efficiency of about 0.27, exhausting to a WHB. The plant meets a major part of the University's heat load of about 7 MW on a mild winter's day. Supplementary firing of the WHB (to about 15 MW) is possible on a cold day. Provision is also made for by-passing the WHB when the heat load is light, in spring and autumn, so that the plant can operate very flexibly, in three modes viz., power only, recuperative and supplementary firing.

The major performance parameters at design operating conditions are as follows:

Electrical power output	3.8 MW
Heat output (normal load)	6.6 MW
(with supplementary firing)	15.0 MW
Gas fuel energy supply	14.95 MW
Thermal efficiency	0.27

Heat/work ratio	1.7
Water supply temperature (T_B)	150°C
Water return temperature (T_A)	128°C
Exhaust gas flow (M_G)	15.3 kg/s
Water flow (M_W)	150 t/h

For WHR operation	EUF = 0.73
($\lambda_{CG} = 1.7$, $\eta_C = 0.4$, $\eta_C = 0.9$)	$\text{FESR} \approx 1 - \dfrac{0.4 \times 0.9}{0.27(0.9 + 1.7 \times 0.4)} = 0.155$

A full description of the economics of operating this plant over a complete year is given by Horlock [1].

References

[1] Horlock, J.H. (1997), Cogeneration—Combined Heat and Power Plants, 2nd edn, Krieger, Malabar, Florida.
[2] Porter, R.W. and Mastanaiah, K. (1982), Thermal-economics analysis of heat-matched industrial cogeneration systems, Energy 7(2), 171–187.

Appendix A

DERIVATION OF REQUIRED COOLING FLOWS

A.1. Introduction

The stagnation temperature and pressure change in the cooling mixing process have been shown to be dependent on the cooling air flow (w_c) as a fraction of the entering gas flow (w_g), i.e. on $\psi = w_c/w_g$. In this Appendix, an analysis by Holland and Thake [1], which allows external film cooling (flow through the blade surface) as well as internal convective cooling (flow through the internal passages), is summarised (see also Horlock et al. [2] for a full discussion). It is based mainly on the assumption that the external Stanton number (St_g), which is generally a weak function of the Reynolds number, remains constant as engine design parameters (T_{cot} and r) are changed.

A.2. Convective cooling only

A simple heat balance for a typical convectively cooled blade (as illustrated in Fig. A.1a, which shows the notation) is

$$Q_{net} = w_c c_{pc}(T_{co} - T_{ci}) = w_g c_{pg}(T_{gi} - T_{go}) = h_g A_{sg}(T_{gi} - T_{bl}). \tag{A1}$$

It is assumed that the temperature of the coolant does not fully reach the temperature of the metal before it leaves the blade, i.e. $T_{co} < T_{bl}$. Thus, the concept of a cooling efficiency is introduced

$$\eta_{cool} = (T_{co} - T_{ci})/(T_{bl} - T_{ci}), \tag{A2}$$

so that

$$Q_{net} = w_c c_{pc} \eta_{cool}(T_{bl} - T_{ci}) = w_g c_{pg}(T_{gi} - T_{go}) = h_g A_{sg}(T_{gi} - T_{bl}). \tag{A3}$$

The exposed area for heat transfer (A_{sg}) is then replaced on the premise that, for a set of similar gas turbines, there is a reasonably constant ratio between A_{sg} and the cross-sectional area of the main hot gas flow A_{xg}. Thus, writing $A_{sg} = \lambda A_{xg} = \lambda w_g/\rho_g V_g$ in Eq. (A3) gives

$$w_c c_{pc} \eta_{cool}(T_{bl} - T_{ci}) = \lambda(h_g/\rho_g V_g)w_g c_{pg}(T_{gi} - T_{bl}),$$

(a) CONVECTIVE COOLING NOTATION

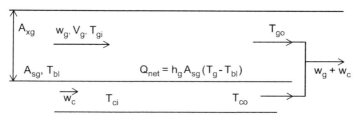

(b) FILM COOLING NOTATION

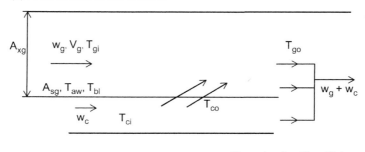

$$Q_{net} = h_{fg} A_{sg} (T_{aw} - T_{bl})$$

Fig. A.1. Notation for turbine blade cooling. (a) Convective cooling and (b) film cooling (after Ref. [2]).

so that

$$(w_c/w_g) = \lambda(c_{pg}/c_{pc})(h_g/c_{pg}\rho_g V_g)(T_{gi} - T_{bl})/\eta_{cool}(T_{bl} - T_{ci})$$

$$= \lambda(c_{pg}/c_{pc})St_g(T_{gi} - T_{bl})/\eta_{cool}(T_{bl} - T_{ci}), \tag{A4}$$

where $St_g = h_g/(c_{pg}\rho_g V_g)$ is the external Stanton number.

For a row in which the blade length is L, the blade chord is c, the spacing is s and the flow discharge angle is α, the ratio λ is given approximately by

$$\lambda = A_{sg}/A_{xg} = 2Lc/(Ls \cos \alpha) = 2c/(s \cos \alpha).$$

With $s/c = 0.8$ and $\alpha = 75°$, the value of λ is then about 10. The total cooled surface area is found to be greater than the surface area of the blade profiles alone because of the presence of cooled end-wall surfaces (adding another 30–40% of surface area), complex trailing edges and other cooled components. It would appear from an examination of practical engines that $\lambda(c_{pg}/c_{pc})$ could reasonably be given a value of about 20. Eq. (A4) then provides the basic form on which a cooling model can be based.

The external Stanton number is assumed not to vary over the range of conditions being studied. Considering $(c_{pg}/c_{pc})(A_{sg}/A_{xg})St_g$ as a constant C, Eq. (A4) then becomes

$$\psi = w_c/w_g = Cw^+ = C\varepsilon_0/\eta_{cool}(1 - \varepsilon_0), \tag{A5}$$

where w^+ is the 'temperature difference ratio' given by

$$w^+ = (T_{gi} - T_{bl})/(T_{bl} - T_{ci}), \tag{A6}$$

and ε_0 is the overall cooling effectiveness, defined as

$$\varepsilon_0 = (T_{gi} - T_{bl})/(T_{gi} - T_{ci}). \tag{A7}$$

T_{gi} and T_{ci} are usually determined from and/or specified for cycle calculation so that the cooling effectiveness ε_0 implicitly becomes a requirement (subject to T_{bl} which again can be assumed for a 'level of technology'). If η_{cool} and C are amalgamated into a single constant K, then

$$\psi = K\varepsilon_0/(1 - \varepsilon_0), \tag{A8}$$

for convective cooling, as used by El-Masri [3].

A.3. Film cooling

The model used by Holland and Thake [1] when film cooling is present is indicated in Fig. A.1b. Cooling air at temperature T_{co} is discharged into the mainstream through the holes in the blade surface to form a cooling film. The heat transferred is now

$$Q_{net} = A_{sg}h_{fg}(T_{aw} - T_{bl}) = w_c c_{pc}(T_{co} - T_{ci}), \tag{A9}$$

where T_{aw} is the adiabatic wall temperature and h_{fg} is the heat transfer coefficient under film cooling conditions. The film cooling effectiveness is defined as

$$\varepsilon_F = (T_{gi} - T_{aw})/(T_{gi} - T_{co}). \tag{A10}$$

Then a new 'temperature difference ratio' W^+ may be written as

$$W^+ = (T_{aw} - T_{bl})/(T_{co} - T_{ci})$$
$$= [\varepsilon_0 - (1 - \eta_{cool})\varepsilon_F - \varepsilon_0\varepsilon_F\eta_{cool}]/\eta_{cool}(1 - \varepsilon_0). \tag{A11}$$

It can be argued that ε_F should be independent of temperature boundary conditions and in the subsequent calculations it is taken as 0.4, based on the experimental data.

It follows from Eqs. (A9) and (A10) that

$$\psi = (w_c/w_g) = (c_{pg}/c_{pc})(A_{sg}St_g/A_{xg})\mu W^+, \tag{A12}$$

where $\mu = h_{fg}/[h_g(1 + B)]$ in which h_{fg} is the heat transfer coefficient under film cooling conditions and $B = h_{fg}t/k$ is the Biot number, which takes account of a thermal barrier coating (TBC) of thickness t and conductivity k.

In practice, h_{fg} increases above h_g, and $(1 + B)$ is increased as TBC is added. For the purposes of cycle calculation, μ is therefore taken as unity and

$$\psi = CW^+, \tag{A13}$$

where C is the same constant as that used for convective cooling only.

A.4. The cooling efficiency

The cooling efficiency can be determined from the internal heat transfer. If T_{bl} is taken to be more or less constant, then it may be shown that

$$\eta_{cool} = 1 - \exp(-\xi), \tag{A14}$$

where $\xi = (h_c A_{sc}/w_c c_{pc}) = (St_c A_{sc}/A_{xc})$, St_c is now the *internal* Stanton number, and A_{sc} and A_{xc} refer to surface and cross-sectional areas of the coolant flow.

Experience gives values of ξ for various geometries, but St_c is also a weak function of Reynolds number and so, in practice, there is relatively little variation in cooling efficiency $(0.6 < \eta_{cool} < 0.8)$. In the cycle calculations described in Chapter 5, η_{cool} was taken as 0.7, and assumed to be constant over the range of cooling flows considered.

A.5. Summary

Since 'open' film cooling is now used in most gas turbines, the form of Eq. (A13) was adopted for the cycle calculations of Chapter 5, i.e.

$$\psi = CW^+. \tag{A15}$$

Taking $(c_{pg}/c_{pc})(A_{sg}/A_g) = 20$ as representative of modern engine practice, and $St_g = 1.5 \times 10^{-3}$ a value of $C = 0.03$ is obtained. The ratio (c_{pg}/c_{pc}) should then increase with T_g (but only by about 8% over the range 1500–2200 K). This variation was, therefore, neglected in the cycle calculations described in Chapter 5.

However, it was found that the cooling flows calculated from these equations were less than those used in recent and current practices in which film cooling is employed. This is for two main reasons:
(i)   designers are conservative, and choose to increase the cooling flows
      (a)   to cope with entry temperature profiles (the maximum temperature being well above the mean) and local hot spots on the blade and
      (b)   locally, where cooling can be achieved with relatively small penalty on mixing loss (and hence on polytropic efficiency), so regions remote from these injection points are cooled with this low loss air;
(ii) in practice, some surfaces in a turbine blade row will be convectively cooled with no film cooling. The use of Eq. (A15) with Eq. (A11) for the whole blade row assembly therefore leads to the total cooling flow being underestimated. Film cooling leads to more efficient cooling, which is reflected in W^+ being much less than w^+; for the NGVs of a modern gas turbine W^+ may take a value of about 2 but w^+ about 4.

In the calculations described in the main text, allowance was made for such practical issues by increasing the value of the constants C by a 'safety factor' of 1.5. Thus, cooling flows were determined from

$$\psi = w_c/w_g = 0.045 W^+, \tag{A16}$$

with

$$W^+ = [\varepsilon_0 - (1 - \eta_{cool})\varepsilon_F - \varepsilon_0\varepsilon_F\eta_{cool}]/\eta_{cool}(1 - \varepsilon_0), \tag{A17}$$

in which ε_F was taken as 0.4 and η_{cool} as 0.7, so that

$$W^+ = [\varepsilon_0 - 0.12 - 0.28\varepsilon_0]/0.7(1 - \varepsilon_0). \tag{A18}$$

In any particular cycle calculation, with the inlet gas temperature T_g known together with the inlet coolant temperature T_{ci}, and with an assumed allowable metal temperature T_{bl}, ε_0 was determined from Eq. (A7). W^+ was then obtained from Eq. (A18) and the cooling flow fraction ψ from Eq. (A16).

References

[1] Holland, M.J. and Thake, T.F. (1980), Rotor blade cooling in high pressure turbines, AIAA J. Aircraft 17(6), 412–418.
[2] Horlock, J.H., Watson, D.E. and Jones, T.V. (2001), Limitations on gas turbine performance imposed by large turbine cooling flows, ASME J. Engng Gas Turbines Power 123(3), 487–494.
[3] El-Masri, M.A. (1987), Exergy analysis of combined cycles: Part 1 Air-cooled Brayton-cycle gas turbines, ASME J. Engng Power Gas Turbines 109, 228–235.

Appendix B

ECONOMICS OF GAS TURBINE PLANTS

B.1. Introduction

The simplest way of assessing the economics of a new power plant is to calculate the unit price of electricity produced by the plant (e.g. $/kWh) and compare it with that of a conventional plant. This is the method adopted by many authors [1,2]. Other methods involving net present values may also be used [3,4].

B.2. Electricity pricing

The method is based on relating electricity price to both the capital related cost and the recurrent cost of production (fuel and maintenance of plant):

$$P_E = \beta C_0 + M + (OM), \tag{B.1}$$

where P_E is the annual cost of the electricity produced (e.g. $ p.a.), C_0 is the capital cost of plant (e.g. $), $\beta(i, N)$ is a capital charge factor which is related to the discount rate (i) on capital and the life of the plant (N years) (see Section B.3 below), M is the annual cost of fuel supplied (e.g. $ p.a.), and (OM) is the annual cost of operation and maintenance (e.g. $ p.a.).

The 'unitised' production cost (say $/kWh) for the plant is

$$Y_E = \frac{P_E}{\dot{W}H} = \frac{\beta C_0}{\dot{W}H} + \frac{M}{\dot{W}H} + \frac{(OM)}{\dot{W}H} \tag{B.2}$$

where $\dot{W}$ is the rating of the plant (kW) and H is the plant utilisation (hours per annum).

The cost of the fuel per annum, M, may be written as the product of the unit cost of fuel ζ($/kWh), the rate of supply of energy in the fuel $\dot{F}$(kW) and the utilisation, H, i.e.

$$M = \zeta \dot{F} H. \tag{B.3}$$

Thus the unitised production cost is

$$Y_E = \frac{P_E}{\dot{W}H} = \frac{\beta C_0}{\dot{W}H} + \frac{\zeta \dot{F} H}{\dot{W}H} + \frac{(OM)}{\dot{W}H} = \frac{\beta C_0}{\dot{W}H} + \frac{\zeta}{(\eta_O)} + \frac{(OM)}{\dot{W}H} \tag{B.4}$$

where $(\eta_O) = \dot{W}/\dot{F}$ is the overall efficiency of the plant. Alternatively, the unit cost of fuel ζ may be written as the cost per unit mass S (say $/kg) divided by the calorific value $[CV]_0$

(kWh/kg), so that

$$Y_{\mathrm{E}} = \frac{\beta C_0}{\dot{W}H} + \frac{S}{[CV]_0(\eta_0)} + \frac{(OM)}{\dot{W}H}. \tag{B.5}$$

In a comparison between two competitive plants, one may have higher efficiency (and hence lower fuel cost) but may incur higher capital and maintenance costs. These effects have to be balanced against each other in the assessment of the relative economic merits of two plants.

B.3. The capital charge factor

The capital charge factor (β) multiplied by the capital cost of the plant (C_0) gives the cost of servicing the total capital required. Suppose the capital costs of a plant at the beginning of the first year is C_0 and the plant has a life of N years so an annual amount must be provided which is ($C_0 i + B$). The first term ($C_0 i$) is the simple interest payment and the second (B) matures into the capital repayment after N years (i.e. interest added to the accumulated sum at the end of each year), thus

$$B[1 + (1 + i) + (1 + i)^2 + \cdots + (1 + i)^{N-1}] = C_0,$$

so that

$$B = \frac{C_0 i}{(1 + i)^N - 1}, \tag{B.6}$$

where it has been assumed that the annual payments are made at the end of each year.
   Hence the total annual payment is

$$C_0 i + B = C_0\left[i + \frac{i}{(1 + i)^N - 1}\right] = C_0 i\left[\frac{(1 + i)^N}{(1 + i)^N - 1}\right] = C_0\beta, \tag{B.7}$$

where the capital charge factor β is sometimes referred to as the annuity present worth factor and is given as

$$\beta = \left[\frac{i(1 + i)^N}{(1 + i)^N - 1}\right].$$

In arriving at an appropriate value of β, the choice of interest or discount rate (i) is crucial. It depends on:

the relative values of equity and debt financing;
whether the debt financing is less than the life of the plant;
tax rates and tax allowances (which vary from one country to another);
inflation rates.

In comparing two engineering projects the practice is often to use a 'test discount rate', applicable to both projects.
   An American approach has been outlined by Williams [1]. He elaborates the simple expression for β to take account of many other factors beyond a simple single interest (or

discount) rate. He defines a discount rate as

$$i = \alpha_e r_e + (1 - \tau)\alpha_d r_d,$$ (B.8)

where α_e, α_d are the fractions of investment from equity and debt, r_e, r_d are the corresponding annual rates of return and τ is the corporate tax rate.

B.4. Examples of electricity pricing

In the unit price of electricity (Y_E) derived in Section B.2, the dominant factors are the capital cost per kilowatt ($C_0/\dot{W}$), which generally decreases inversely as the square root of the power (i.e. as $\dot{W}^{1/2}$), the fuel price ζ, the overall efficiency η_O, the utilisation (H hours per year) and to a lesser extent the operational and maintenance costs (OM).

Fig. B.1 shows simply how Y_E, minus the (OM)/$\dot{W}H$ component, varies with $C_0/\dot{W}$ and η_O, for $H = 4000$ h and $\zeta = 1$ c/kWh. Horlock [4] has used this type of chart to compare three lines of development in gas turbine power generation:

(i) a heavy-duty simple cycle gas turbine, of moderate capital cost, with a relatively low pressure ratio and modest thermal efficiency (e.g. 36%);

(ii) an aero-engine derivative simple cycle gas turbine, usually two-shaft and of high pressure ratio, the capital cost per kilowatt of this plant being surprisingly little different from (i) in spite of it being derived from developed aero-engines, but thermal efficiency being slightly higher (e.g. 39%);

(iii) a heavy-duty CCGT plant, based on (i), which has a high thermal efficiency but increased capital cost.

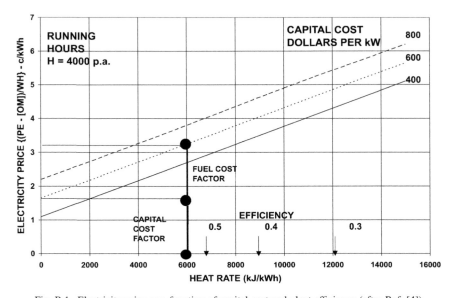

Fig. B.1. Electricity price as a function of capital cost and plant efficiency (after Ref. [4]).

Rough locations for types (i), (ii) and (iii) are given in the electricity price charts of Figs. B.2 and B.3; for 8000 and 4000 h utilisation, respectively. For 8000 h, the CCGT plant type (iii) has a clear advantage in spite of increased capital costs. At 4000 h, the CCGT plant loses this advantage over the aero-engine derivatives because of the increase in the capital cost element (H has been decreased).

However, more direct comparisons should include factors of operation and main-tenance, the cost of which have been omitted in the presentations of Figs. B.2 and B.3.

B.5. Carbon dioxide production and the effects of a carbon tax

As pointed out in Chapter 7, the amount of CO_2 produced by a thermal plant is now a major criterion of its performance, for environmental and therefore economic reasons.

In electrical power stations a new measure of the performance is the amount of CO_2 produced per unit of electricity generated, i.e. $\lambda = kg(CO_2)/kWh$; this quantity can be non-dimensionalised by writing $\lambda' = \lambda(16/44)(LCV)$ where $(16/44)$ is the mass ratio of fuel to CO_2 for methane and (LCV) in its lower heating value. However, presenting the plant's 'green' performance in terms of λ directly allows the cost of any tax on the carbon dioxide to be added to the untaxed cost of electricity production most easily.

Fig. B.4 (after Davidson and Keeley [5]) shows values of λ plotted against thermal efficiency for a high carbon fuel (coal) and a lower carbon fuel (natural gas). It illustrates that one obvious route towards a desired low production of this greenhouse gas is to seek high thermal efficiency (another is to use lower carbon fuel).

In future, the economics of electric power generation is likely to be affected considerably by the amount of CO_2 produced and the level of any environmental penalty

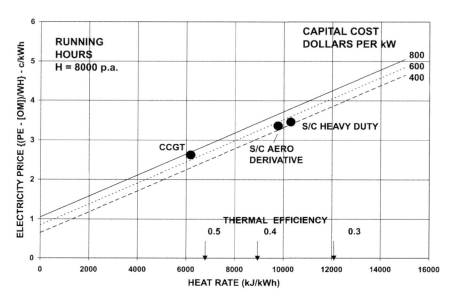

Fig. B.2. Electricity price for typical gas turbine plants—running hours 8000 p.a. (after Ref. [4]).

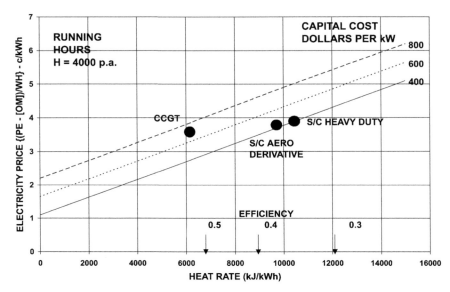

Fig. B.3. Electricity price for typical gas turbine plants—running hours 4000 p.a. (after Ref. [4]).

imposed by a carbon or carbon dioxide tax. For example, a CCGT plant of 54% thermal efficiency, delivering electricity at a generating cost of 3.6 c/kWh can produce CO_2 at a rate of 0.3 kg/kWh, as indicated in Fig. B.5. If the carbon dioxide tax is set at \$50/tonne of CO_2 (5 c/kg CO_2), then there is an additional amount of $(0.3 \times 5) = 1.5$ c/kWh to be

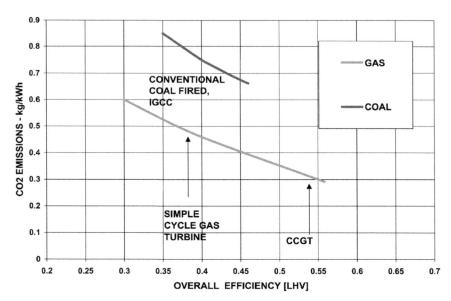

Fig. B.4. Carbon dioxide emissions for various power plants as a function of overall efficiency (after Davidson and Keeley [5]).

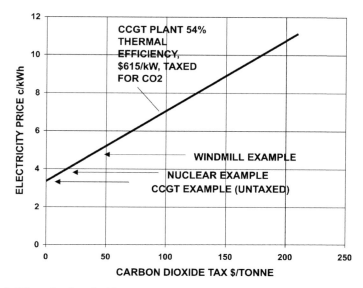

Fig. B.5. Effect of carbon dioxide tax on electricity price for a combined cycle gas turbine plant.

added to the cost of generation, making it 5.1 c/kWh. This may make the plant uneconomic when compared to a nuclear station or even windmills. This point is illustrated in Fig. B.5 which shows how the generation cost for this CCGT plant would vary with the tax level and how other plants might then come into competition with it.

If however, the original CCGT plant was modified to reduce the amount of CO_2 entering the atmosphere from the plant (say to 0.15 kg/kWh) at an additional capital cost it may lead to an increase in the untaxed cost of electricity (say from 3.6 to 4.2 c/kWh). Then the effect of a carbon dioxide tax of 5 c/kWh would be to increase the electricity price to $(4.2 + 0.15 \times 5) = 4.95$ c/kWh and this is below the 'taxed' cost of the original plant. In fact, the new plant would become economic with a carbon dioxide tax of T c/kg CO_2, which is given as $(3.6 + T \times 0.3) = (4.2 + T \times 0.15)$, i.e. when $T = 4$ c/kg CO_2.

References

[1] Williams, R.H. (1978), Industrial Cogeneration, Annual Review of Energy 3, 313–356.
[2] Wunsch, A. (1985), Highest efficiencies possible by converting gas turbine plants, Brown Boveri Review 1, 455–456.
[3] Horlock, J.H. (1997), Cogeneration—Combined Heat and Power Plants, 2nd edition, Krieger, Malabar, Florida.
[4] Horlock, J.H. (1997), Aero-engine derivative gas turbines for power generation: thermodynamic and economic perspectives, ASME Journal of Engineering for Gas Turbines and Power 119(1), 119–123.
[5] Davidson, B.J. and Keeley, K.R. (1991), The thermodynamics of practical combined cycles, Proc. Instn. Mech. Engrs., Conference on Combined Cycle Gas Turbines, 28–50.

SUBJECT INDEX

ABB GT24/36 CCGT plant, 128
Absorption, 136–139
Adiabatic combustion, 23
Adiabatic mixing, 51
Adiabatic wall temperature, 185
Advanced steam topping (FAST), 99, 100
Aero-engine derivative, 191
Aftercooler, 94–96
Air recuperation, 90
Air standard cycles, 28, 33, 48, 68
Allowable stack temperature, 118, 174
Ambient temperatures, 13–14, 24
Annual cost, 189
Annual payments, 190
Annuity present worth factor, 190
Arbitrary overall efficiency, 6–7, 40–42, 66, 112–113, 168
Area for heat transfer, 183
Area plots of the range of EUF and FESR, 179
Artificial efficiency, 170
Arbitrary overall efficiency, 41
Artificial thermal efficiency, 170

Basic power plant, 2
Basic STIG plant, 85
Basic gas turbine cycles, 27–46
Beilen CHP plant, 177, 180
Biot number, 185
Bled steam feed water heating, 119–120, 121
Boiler efficiency, 5, 111, 117
Boiler pressure, 118
Boudouard reaction, 143

Calculated exergy losses, 83
Calculating plant efficiency, 71–84
Calorific value experiment, 5, 14, 41, 87, 90
Calorific value, 5–6, 14, 41, 87, 90, 189–190

Capital charge factor, 189, 190–191
Capital cost per kilowatt, 191
Capital costs, 131, 132, 189, 190–192
Carbon dioxide, 131, 192, 193
Carbon dioxide removal, 144–145, 146, 157
Carbon tax, 163–164, 192–194
Carnot cycle, 7, 8, 9, 20
Carnot efficiency, 7, 9
Carnot engines, 7–9, 16–17, 20
Cascaded humid air turbine (CHAT) cycle, 101, 102, 104, 107
CBT and CCGT plants with full oxidation, 158
CBT open circuit plant, 39
CCGT (combined cycle gas turbines), xiv, 109, 111, 112, 116, 117, 123
CCGT plant with feed water heating by bled steam, 119
CCGT plant with full oxygenation, 158
Change in overall efficiency, 21–22, 127
Change in total pressure, 62
Centrax 4 MW gas turbine, 180
CHAT (cascaded humid air turbine) plant, 101, 102, 104, 107
Chemical absorption, 137
Chemical absorption process, 137
Chemical reactions, 22, 141–145
   reforming, 143, 148, 157
Chemically reformed gas turbines (CRGT), 133, 147–153
CHP see combined heat and power
CHP plant, 3, 167, 174, 177
Classification of gas-fired plants, 132
Classification, gas-fired cycles, 132–136
Closed circuit gas turbine plant, 2, 4
Closed cyclic power plant, 1
Closed cycles
   air standard, 33
   efficiency, 4–6
   exergy flux, 19–22

power generation, 1
steady-flow energy equation, 13
CO_2 produced per unit of electricity, 192
CO_2 removal at high pressure level, 135
CO_2 removal at low pressure level, 135
CO_2 removal equipment, 136
Coal fired IGCC, 115, 164
Cogeneration plant, 3, 4, 167, 168
Cogeneration plants *see* combined heat and
      power plants
Combined cycle gas turbines (CCGT), 109,
      112–129
Combined heat and power plant, 3, 167, 174,
      177
Combined power plant, 2, 4, 109
Combined STIG cycle, 99
Combined heat and power (CHP) plants
      operation ranges, 174–177
      performance criteria, 168–173
      power generation, 1
      unmatched gas turbines, 173–174, 175
Combined heat and power (CHP) plants, xi,
      167–181
Combined plants, 109–113
      efficiency, 111
      power generation, 1
      steam injection turbines, 99
      see also combined cycle gas turbines;
            combined heat and power
Combustion temperature, 48, 56
Combustion with fuel modification, 160
Combustion with full oxidation, 160
Combustion with recycled flue gas, 144
Combustion with excess air, 141
Combustion
      complete, 140–141
      fuel modification, 133, 134, 147–153
      open circuit plants, 39–42
      oxidant modification, 135, 154–161
      recycled flue gases, 144
      temperatures, 47–57, 65–68, 73–81
Combustor outlet temperature, 47
Complete combustion, 140–141
Completely dead state, 22
Complex cycle with partial oxidation and
      reforming, 157
Complex RWI cycles, 105
Component performances, 33–34
Compressor water injection, 101–102

Computer calculations, 43–45, 65–68, 75–81
Constant pressure closed cycle
      see Joule–Brayton cycle
Convective cooling, 71–72, 183–185
Conventional power plant, 1
Cool Water IGCC plant, 115
Cool Water pilot plant, 114
Coolant air fractions, 74, 79
Cooled efficiency, 56, 58
Cooling air flow, 183
Cooling
      air fractions, 57, 65, 71–84, 184–187
      air-standard cycles, 48–55, 51, 54–59
      effectiveness, 185
      efficiency, 72–73, 183, 186
      flow fraction, 60, 65, 187
      flows, 47–68, 71–73, 183–187
      mixing processes, 183
      plant efficiency, 71–73
      reversible cycles, 49–54
      thermal efficiency, 47–68
      turbine blade rows, 59–65, 186
Cooling of internally reversible cycles, 49
Cooling of irreversible cycles, 55
Corporate tax rate, 191
Cost of electricity, 131, 163
Costs, 131, 132, 190–192
CRGT (chemically reformed gas turbines), 133,
      148–153
Cycle...
      analysis parameters, 8–9, 20–21
      calculations, 65–68
      efficiency *see* thermal efficiency
      widening, 9, 21
Cycles burning non-carbon fuel (hydrogen),
      152
Cycles with modification of the oxidant in
      combustion, 154
Cycles with perfect recuperation, 92

Dead state, 15, 22
Debt financing, 190
Delivery work, 22
Demand loads, 170–173
Derivation of required cooling flows, 183–187
Design, combined heat and power plants, 177
Development of the gas turbine, xi
Dewpoint temperature, 114, 119, 122
Direct removal of CO_2, 145

Direct removal, carbon dioxide, 144–145
Direct water injection cycles, 103
Discount rate, 190–191
Disposal, carbon dioxide, 132
Dry and wet cycles, 104
Dry efficiency, 94
Dry recuperative cycles, 91
Dual pressure systems, 121, 123, 129
Dual pressure system with no low pressure water
    economiser, 123
Dual pressure system with a low pressure
    economiser, 123

Economic viability, 163
Economics of a new power plant, 189
Economics, 131, 132, 163–164, 189–194
Economiser water entry temperature, 119, 120
Effect of carbon dioxide, 194
Effect of steam air ratio, 89
Effectiveness (or thermal ratio), 33
Efficiency, 4
Efficiency
    closed circuit plants, 4–6
    combined cycle turbines, 126
    dry, 94
    exhaust heated combined cycles, 112–114
    fired combined cycle turbines, 116
    Joule–Brayton cycle, 1, 3, 9, 10, 20, 28
    maximum, 35, 38, 66, 81, 126
    open circuit power plants, 6–7
    plants, 71–84
    power generation, 9
    rational, 6, 22, 24–25
    steam injection turbine, 87–89
    water injection evaporative turbines, 94–98
    see also plant efficiency; thermal efficiency
EGT *see* evaporative gas turbines
Electricity pricing, 131, 163–164, 189–192
El-Masri EGT cycles, 96
End of pipe CO_2 removal, 132, 164
Energy equations, 13, 85, 87, 91, 172
Energy utilisation factor (EUF), 7, 168–169,
    174–177, 178–179
Enthalpy, 13–14, 33–34
    changes, 43, 61–62
    entropy diagrams, 91–92
    flux, 13, 90
    specific, 24
    steam, 119–120, 121

Entropy, 9, 16–17, 24, 64–65, 91–92
    see also temperature–entropy diagrams
Entropy generation, 65
Entry feed water temperature, 119, 120
Equilibrium constants, 143
Equipment to remove carbon dioxide, 132
Equity and debt financing, 190
EUF *see* energy utilisation factor
Evaporative gas turbines (EGT), 85, 91–98,
    99–102
Exergy flux, 19
Exergy losses, 25, 83
Exergy, 13, 15, 82–83
    equation, 23
    flux, 19–21, 23, 25
    losses, 83–84, 100–102
Exhaust, 112–14, 116–122, 140–141
Exhaust heated (supplementary fired) CCGT, 116
Exhaust heated (unfired) CCGT, 112
Exhaust irreversibility, 14, 19, 83
Exit turbine temperature, 59
External irreversibilities, 8
External Stanton number, 184–185
Extraction work, 22

FAST cycle, 99, 103
Feed heating, 114, 116, 119–123, 128, 129
Feed water temperature, 114, 120, 122, 123
FESR, 171, 172, 173, 174, 176, 177, 180, 181
    see fuel energy saving ratio
(FG/TCR) cycle, 152
    see Flue Gas thermo-chemical recuperation
Film cooling, 72–73, 183, 184, 185
Fired combined cycle gas turbines, 116–123,
    174–177
First industrial gas turbine, xiii
Flows
    cooling, 47–68, 71–73, 183–187
    mainstream, 71–72
    mass flow, 42, 71–72, 117–118
    work, 14–18
    see also steady-flow
Flue Gas thermo-chemical recuperation
    (FG/TCR), 133, 144–145, 151–153
Fluid mechanics, 59–65
Foster–Pegg plant, 99
Fuel...
    air ratio, 41–42

energy saving ratio (FESR), 170–177,
    179–180
  modification, 133–135, 147–152
  per annum costs, 189
  price, 191
  saving, 170–173
Full oxidation, 134–135, 158–160

Gas supplied for combustion, 150
Gas turbine jet propulsion, xiii
Gas turbine, xiii
Gaseous fuel, 23
Gasifier, 114
GEM9001H plant, 128
General electric LM 2500 [CBT] plant, 83
General Electric company, 114
Gibbs function, 22
Graphical method, 35–36, 123–125
Global warming, 131
Greenhouse gases, 131
  see also carbon dioxide removal
Gross entropy generation, 64–65

HAT cycle, 100, 106
HAT *see* humidified air turbines
Heat balance in the HRSG, 118
Heat...
  balance, 90, 118–119, 183
  electrical demand ratio, 170–173, 176–177
  engines *see* closed cycles/circuits
  exchange (or recuperation), 10, 91–92,
      94–98, 133, 147–150
  exchanger, 11, 32, 96
  exchanger effectiveness, 37, 93
  loads, 170–174
  loss in the exhaust stack, 172
  loss, 110–112
  rate, 7
  recovery steam generator (HRSG), 85
    combined cycle gas turbines, 112,
        114–115, 118–121, 126–128
    combined heat and power plants, 180
    steam injection turbine plants, 87–88
  rejection, 8–9, 18
  supply, 8–9, 37
  transfer, 5, 14–17, 183–185, 186
  transfer coefficient, 185
  to work ratio, 175, 176–177, 179, 180
Heating device (or boiler) efficiency, 5, 111, 117

Heating value, 143, 150, 152
Heavy duty CCGT plant, 191
Heat Recovery Steam Generator HRSG, 112,
    114, 116
Humidified air turbine, 100, 101, 104
Hydrogen burning CBT, 133
Hydrogen burning CCGT, 133, 154
Hydrogen plants, 133, 153–154

ICAR (irreversible Carnot), 22
Ideal (Carnot) power plant, 7–8
Ideal combined cycle plants, 109–110
Ideal heat exchangers, 91
IFB plant, 103
IFB *see* inlet fog boosting
IGCC cycles with CO_2 removal, 160
IGCC *see* integrated coal gasification cycles
Integrated coal gasification combined cycle
    plant (IGCC), 114, 115
IJB *see* irreversible Joule–Brayton
Inlet fog boosting (IFB), 103
Integrated coal gasification cycles (IGCC),
    114–115, 136, 161–162, 164
Intercooled cycle, 32, 96
Intercooling and reheating, 39, 93
Intercooled steam injection turbine plants
    (ISTIG), 97–98, 103, 105
Intercooling, 10–11
Interest rates, 190–191
Internal irreversibilities, 8–9, 16, 19, 24
Internal irreversibility, 16, 19, 24
Internal Stanton number, 186
Internal thermal efficiency, 50
Internally reversible cycles, cooling, 49–55
Irreversible Carnot (ICAR) cycles, 22
Irreversible Joule–Brayton (IJB) cycle, 9, 21
Irreversible processes
  air standard cycles, 33–39, 51, 54–59
  power generation, 8–9
  steady flow, 14, 17–18
Irreversibility, 14, 17
Irreversible Joule-Brayton (IJB) cycle, 9, 20
Irreversible simple cycle, 34
Isentropic efficiency, 33
Isentropic...
  efficiency, 33–34
  expansion, 53–54
  temperature ratio, 35–39, 43, 66–67,
      92–93

ISO firing temperature, 47
Isothermal compression, 93
ISTIG plant, 98, 103, 105
   see intercooled steam injection turbine
      plants

Joint heating of gas turbine and steam turbine
      plants, 112
Joule–Brayton cycle, 1, 3, 20, 28
Joule–Brayton (JB) cycle
   air standard, 28–29, 46
   efficiency, 9, 10
   exergy flux, 20–22
   power generation 1–2, 3

Linearised analyses, 42
Liquefaction, 134
Liquid fuel, 23
Live steam pressure, 122
Liverpool University plant (CHP), 180–181
Loss in efficiency, 58, 110
Lost work, 16, 17–18, 20–21
Lower heating value thermal efficiency, 124

Mach numbers, 62
Mainstream gas mass flow, 71–72
Maintenance costs, 191
Mass flow, 42, 71, 117–118
Mass flow ratio, 118
Matched CHP plant with WHB, 171
Matched CHP plant with WHR, 171
Matched plants, 171
Matiant cycle, 134–135, 158–160
Maximum combined cycle efficiency, 126
Maximum efficiency, 35, 38, 66, 82, 126
Maximum efficiency, 126
Maximum (reversible) work, 17
Maximum specific work, 35
Maximum work, 15, 22
Maximum work output, 22, 24–25
Maximum temperature, 47
Mean temperatures 8–9, 21
Methane, 141–143, 145, 192
Mixing of cooling air with mainstream flow, 61
Modifications
   fuels, 133–135, 148–153
   oxidants, 134–135, 155–161
   turbine cycles, 9–11
Modified polytropic efficiency, 59

Multi-step cooling, 52–54, 59, 75, 78–81
Multiple PO combustion plant, 163

Natural gas reforming, 133–134
Natural gas-fired plants, 164
NDCW *see* non-dimensional compressor work
NDHT *see* non-dimensional heat transferred
NDNW *see* non-dimensional net work
NDTW *see* non-dimensional turbine work
Nitrogen, 133, 153
Non-carbon fuel plants, 133, 153–155
Non-dimensional heat supplied, 41
Non-dimensional net work output, 40
Non-dimensional...
   compressor work (NDCW), 35, 124
   heat transferred (NDHT), 3, 122
   net work (NDNW), 35–37, 40, 123
   turbine work (NDTW), 35, 124
Notation, turbine cooling, 184
Novel gas turbine cycles, 131–164
Nozzle guide vane rows, 60, 63, 65, 73–75, 78

Open circuit gas turbine plant, 2, 6, 13, 24, 39,
      43
Open circuit gas turbine/closed steam cycle, 113
Open cooled blade row, 61, 62
Open cooling, 59–65, 186
Operating conditions/ranges, 180–181
Operational costs, 191–192
Operation and maintenance, 192
Optimum pressure ratios, 44–45, 123–126
Overall cooling effectiveness, 185
Overall efficiency and specific work, 66, 78, 81
Overall efficiency of CCGT plant, 121, 124
Overall efficiency
   closed circuit power plants, 6
   cogeneration plants, 167–169
   combined cycles, 112, 118, 128, 129, 130
   electricity pricing, 189–190
   fired combined cycles, 116
   open circuit plants, 43–46
   open circuit power plants, 6–7
   recuperation, 92, 149–151
   steam injection turbine plants, 85, 86
   steam-thermo-chemical recuperation, 33,
      141, 143, 147
   three step cooling, 79–81
   water injection evaporative turbines,
      94–98

wet gas turbine plants, 85, 87–107
see also arbitrary...
Oxidant modification, 135, 163
Oxygen blown integrated coal gasification
        cycles, 161, 162

Parallel expansions, 51
Parametric calculations, 118–121
Parametric studies, 97, 105, 107
Partial oxidation (PO), 134–135, 143, 155–157
Partial oxidation cycles, 155
Partial oxidation reaction, 143
Performance criteria, 33, 168
Performance of unmatched CHP plants, 175
Physical absorption process, 136, 138
Physical absorption, 137, 139–140
Pinch point temperature difference, 88, 118
Plant with a WHB, 174
Plant with supplementary firing, 116
Plants with combustion modification, 158
PO open CBT cycle, 135
PO plant with CO_2 removal, 157
PO, 141, 143, 154, 155
Plant efficiency
    calculations, 71–83
    electricity pricing, 189, 191–194
    exergy, 82–83
    turbine cooling, 68
PO *see* partial oxidation
Polytropic efficiency, 34, 59, 64
Polytropic expansion, 53, 59
Power
    generation thermodynamics, 1–11
    loads, 173–174
    plant performance criteria, 4
    station applications, 131
Practical gas turbine cogeneration plants, 177
Pre-heating loops, 122–123
Pressure
    change, 62
    dual systems, 123
    live steam, 122–123
    losses, 33, 39, 75, 78
    ratios
        optimum, 44–45, 123–126
        turbine cooling, 66–68
        water injection evaporative gas turbines,
            96–98

stagnation, 60, 61–65, 183
steam raising, 119–120, 121
two step cooling, 51–52
Process steam temperatures, 177, 178
Product of thermal efficiency and boiler
        efficiency, 6, 111

Range of EUF and FESR, 177, 179
Range of operation, 174
Rankine type cycles, 133, 154–155
Ratio of entropy change, 9
Rational efficiency, 6, 22, 24–26, 42,
        51, 60
Rayleigh process, 62
Real gas effects, 39, 43, 45, 46, 48, 65,
        71, 82
Recirculating exhaust gases, 140–141
Recuperated water injection (RWI) plant,
        100–101, 104, 106–107
Recuperation (heat exchange), 10–11, 90–92,
        133, 147–150
Recuperative CBTX plant, 147
Recuperative cycle, 29, 30, 34, 37, 38, 92
Recuperative STIG plant, 90
Recuperative STIG type cycles, 148
Recycled flue gases, 144
Reference systems, 170–173
Reforming reactions, 143, 148, 157, 158–159
Regenerative feed heating, 116, 122, 128
Reheat and intercooling, 10, 11
Reheating in the upper gas turbine, 126
Reheating, 31, 39, 44, 45, 46, 126–128
Rejection, heat 8–9, 18
REVAP cycle, wet gas turbine plants, 100–101,
        104, 108
Reversed Carnot engine, 18
Reversibility and availability, 13–26
Reversible closed recuperative cycle, 30
Reversible processes
    air standard cycles, 28–33, 46, 49
    ambient temperature, 14–15
    availability, 13–26
    heat transfer, 15–17
Reynolds number, 183, 186
Rolls-Royce, plc, xiii–xv, 83–84
Rotor inlet temperatures, 47–54, 56–57, 60,
        65–68
Running costs, 131

Ruston TB gas turbine, 177, 180
RWI cycle, 100, 101, 103, 105, 106
RWI *see* recuperated water injection

Safety factor (cooling), 186
Scrubbing process, 147–148
Semi-closure cycles, 134, 140–141, 146–148, 157, 159–162
Semi-closed CBT or CCGT, 134
Semi-closed CCGT plant with CO_2 removal, 163, 164
Semi-closed CICBTBTX cycle, 135
Semi-closure, 139, 140, 158
Sequestration, 132, 134, 145–148
Shift reactor, 161–162
Simple CHT cycle, 34
Simple EGT, 93, 96, 107
Simple PO plant, 155
Single pressure system, 122–123
Simple single pressure system with feed heating, 122
Simple single pressure system without feed heating, 118
Single pressure steam cycle with LP evaporator in a pre-heating loop, 123
Single pressure steam raising, 121
Single-step turbine cooling, 49–51, 55–57, 73–75, 76–78
Specific enthalpy, 24
Specific entropy, 24
Specific heat, 35, 41–42, 43, 88
Specific work
    closed air standard cycles, 35
    combined cycles, 123–124
    open circuit plants, 45–46
    steam-thermo-chemical recuperation, 150, 151
    wet gas turbine plants, 104–107
Stack temperature, 119
Stagnation pressure/temperature, 60, 61–65, 183
Stanton numbers, 183, 184–185, 186
Stationary entry nozzle guide vane row, 60–65
Steady-flow, 1, 13
    availability function, 14, 15, 23, 24
    energy equation, 13, 85, 87, 91, 172
Steam
    air ratios, 87–89, 150
    enthalpy, 119

injection turbine plants (STIG), 85–86
    intercooled, 97–98, 103, 105
    recuperation, 91–94, 133, 149–150
    thermodynamics, 103
reforming reactions, 143, 144, 148
thermo-chemical recuperation, 133, 143, 149, 150
turbines, 128
Steam cooling of the gas turbine, 128
Steam injection and water injection plants, 86
STIG and EGT, 85, 97, 103
STIG cycle, 96, 97, 99, 103, 107
Stoichiometric limit, 47
STIG *see* steam injection turbine plants
Sulphuric acid dewpoint, 122
Supplementary combustion, 172
Supplementary firing, 116, 173
Supplementary fired CHP plant, 172
Supplementary 'heat supplied', 120
Surface intercoolers, 105
Syngas, 114–115, 136, 143–144, 161–162

Taxes, 131, 162–164, 191, 192–194
Tax rates, 190
TBC (Thermal barrier coating), 185
TCR, 133, 141–143, 147–152, 157
TCR *see* thermo-chemical recuperation
Temperature
    adiabatic wall, 185
    ambient, 13–14, 24
    changes, 39, 42–43
    combustion, 47–49, 55–57, 68, 73–84
    dewpoint, 114, 119, 122
    difference ratio, 71–72, 185, 187
    economiser water entry, 119
    exit turbine, 59
    isentropic ratio, 35–39, 43, 66–67, 92–93
    ISO firing, 47
    mean, 8, 21
    pinch point, 118
    power generation, 8–9
    process steam, 177, 178
    rotor inlet, 47–54, 56–57, 65–68
    stack, 118
    stagnation, 60, 61–65, 183
    turbine entry, 50, 58

Temperature–entropy diagrams
  air standard cycles, 28, 33
  combined cycle efficiency, 117
  evaporative gas turbines, 91, 92
  fired combined cycles, 116
  ideal (Carnot) power plants, 7
  intercooling, 32–33
  Joule–Brayton cycles, 1, 3, 28
  multi-step cooling, 52
  single-step cooling, 49–50, 55
  thermal efficiency, 6–11
  two-step cooling, 51, 58
  water injection evaporative gas turbines,
    94–96
Temperature–entropy diagrams, xiv
Texaco gasifier, 114
Thermal barrier coating (TBC), 185
Thermal efficiency
  air standard cycles, 30–31, 35–37
  artificial, 168
  closed circuit power plants, 3–6
  combined heat and power plants, 110–111,
    168
  cooling flow rates, 47–68
  evaporative gas turbines, 85
  fired combined cycles, 117–126
  ideal (Carnot) power plants, 7
  ideal combined cyclic plants, 109–110
  internal, 50
  irreversible Joule–Brayton cycle, 20
  modifying turbine cycles, 9–11
  open circuit power plants, 6
  recuperative evaporative gas turbines,
    92–93
  steam injection turbine plants, 89
  three step cooling, 79, 81
  turbine cooling, 47–68
Thermal energy, 18, 24
Thermal or cycle efficiency, 5, 7
Thermal ratio, 33
Thermo-chemical recuperation (TCR), 133, 134,
    142–144, 148–153
Thermodynamics
  open cooling, 59–65
  power generation, 1–11
  wet gas turbine plants, 103–105
Three step cooling, 78–79, 80–81
Throttling, 52, 58
TOPHAT cycle, 101–102, 104, 107

Total pressure loss, 63–65
Turbine
  cooling, 47–69, 184, 186–187
  entry temperature, 47, 50, 56, 58, 119
  exit condition, 54–55
  mass flow, 42
  pressure, 157–158
  work, 88, 94, 96
Turbo jet engines, xiii
Two pressure systems, 121, 123, 129
Two-step cooling, 51–52, 58

Ultimate reversible gas turbine cycle, 33
Uncooled and cooled efficiencies, 57
Unfired plant, 112–114, 167, 170,
    174–177
Unit costs, 189
Unit price of electricity, 189, 191–192
Unitised production costs, 189
Unmatched gas turbines, 173–174, 175
Unused heat, 110, 176–177
Upper gas turbine cycles, 126–128
Useful heat/work, 177, 178

Value-weighted energy utilisation factor, 169
Van Liere cycle, 92, 101–102, 107
Van't Hoff box, 142, 143

Waste heat boilers (WHB), 167–177, 180
Waste heat recuperators (WHR), 167–77,
    180–181
Water
  entry temperature, 114, 119, 122
  gas shift reactions, 142–144
  injection, 85–107
    evaporative gas turbines, 94–98
Water injection into aftercooler, 95
Water injection into aftercooler and cold side of
    heat exchanger, 95
Water injection into cold side of heat exchanger,
    95
Westinghouse, 83–84
Westinghouse/Rolls-Royce WR21 recuperated
    [CICBTX]$_I$ plant, 83
Wet and dry cycles compared, 104, 105
Wet efficiencies, 94
Wet gas turbine plants, 85–107

WHB *see* waste heat boilers
Whittle laboratory, xv
WHR *see* waste heat recuperators
Work
   irreversible flow, 15, 17
   lost, 16, 17–18, 20–21

open circuit plants, 39–42
output, 22, 24–26
potential, 18, 19, 24
reversible flow, 14, 16
turbine, 88, 94, 96
see also specific work

Printed and bound by CPI Group (UK) Ltd, Croydon, CR0 4YY

08/05/2025

01864854-0003